A Culinary Voyage around the Greek Islands

THEODORE KYRIAKOU

A Culinary Voyage around the Greek Islands

THEODORE KYRIAKOU

Photography by Jason Lowe

Quadrille

For Freddie, my friends, my relations and my mother
for much inspiration, and the endless gastronomic moments
while we were crossing the misty blue horizon.

First published in 2008 by Quadrille Publishing Limited,
Alhambra House, 27-31 Charing Cross Road,
London WC2H OLS

Editorial Director: Jane O'Shea
Creative Director: Helen Lewis
Editor and Project Manager: Lewis Esson
Art Director: Lucy Gowans
Photography: Jason Lowe
Production: Ruth Deary

Cataloguing in Publication Data: a catalogue record for this book is available from
the British Library.

ISBN: 978 184400 604 5

Printed and bound in China

Contents

A Short Prologue 6

CHAPTER 1
At The Mercy of the First Breakfast 8

CHAPTER 2
Spring Tides, Full Moon and Seasonal Vegetables 30

CHAPTER 3
Perfidious Greek Herbs 58

CHAPTER 4
Crustaceans Drowning in a Greek Coffee Pot Full of
Aegean Water 82

CHAPTER 5
Cooking While Heading Into a Thunderstorm 112

CHAPTER 6
Meats That Melt in the Mouth 134

CHAPTER 7
Dinner Time Under the Night Firmament 150

CHAPTER 8
After the Big Waves, We Are in Every Sense Very Hungry 172

CHAPTER 9
Sugar and Honey Make the Storm Lantern in our Hearts 194

A Brief Guide to the Greek Islands and their Food 218
The Greek Island-hopping Calendar 219
Glossary of Greek Ingredients 220
Index 221
Acknowledgements 223

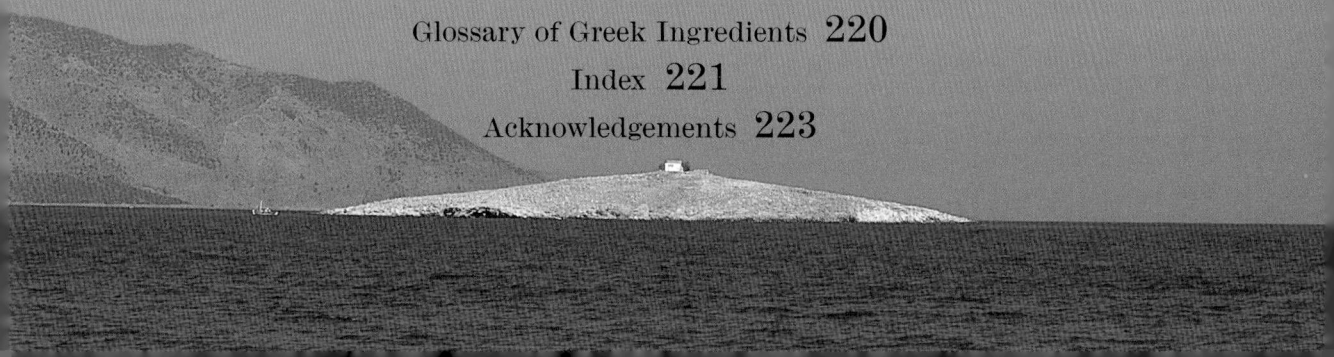

A Short Prologue

'To travel hopefully is a better thing than to arrive...'

Robert Louis Stephenson

You remove your shoes before you come on board and, as you drop your minimalist luggage, you find yourself wearing just swimming trunks, looking at the harbour disappearing into the distance, while a delicate smell of freshly chopped dill is wafting from the open galley as it falls like confetti into the bowl of sea urchins with olive oil and lemon juice.

When on or around the Greek Islands, I never feel like pulling blinds or closing curtains, turning on air-con or staging an exit. From a mere whiff out of the kitchen, you are able to guess how much longer it will be before the table is set.

Island food has an honesty about it and that is what makes the Greek Islands and sailing around them quite so appealing. The Islands and the deep blue Aegean waters are a great place to develop a cuisine. The natives and sailors are invariably forced to work with what to the mainlander might seem like a restricted palette, and island cuisine is more likely to be seasonal. Great skill and simplicity is needed to network the same ingredients and themes repeatedly while keeping them fresh.

In the last few years I have been lucky enough to sail round the Islands regularly in the summer while teaching groups of people how to cook Greek food. These trips have been wonderful experiences, and not just because several times I have been able to take along those closest to me and introduce them to this wonderful blue paradise – including my 'spiritual wife' Chantal, my business partner Paloma and her little son, my godson Freddie.

Recently we also made a special trip with the team working on the book to photograph the food. Having to source my ingredients on the Islands in this way has deepened my understanding and appreciation of the Islanders simple but glorious way of life and their ingenuity with what and how they cook.

In this book, I return to the environment that makes me feel at home, the Greek Islands, the sea and the boat. I feel like a voyeur with culinary Aegean island tendencies. The book is divided into different food according to various times of the day, weather conditions, dinners while star-gazing, spring ingredients, the particular smell of the land, Easter's full moon, diving for shellfish and sea urchins, buying milk-fed lamb or kids while visiting local festivals, island gardens and the limited but unusual desserts. I hope you will enjoy them all as much as we have.

Theodore

1

At the Mercy of the First Breakfast

Water supply on the majority of the Greek islands or on board a boat is very limited, especially during the midsummer months. Water is either brought in barges or collected while it is raining, so conservation is essential. Before breakfast, we use the sea rather than the shower for waking up.

The surest way to our early morning hearts is through our stomachs. Whether you are sitting on a veranda or have been tightly packed around a small round table with no menus, breakfast has always been a solemn ritual, where simple mouthwatering flavours become hopelessly and irreversibly addictive, while sipping good Greek coffee.

Breakfast in Greece is often more savoury than sweet, so the olive plays its role, together with warm bread, not-too-salty cheeses like fresh anthotiro or mizithra, throumbes olives, fried eggs drizzled with lemon juice, coffee of course or black tea with a dollop of honey and lemon, a steaming plate of trahana porridge, and often a shot glass of tsipouro (grappa). Other usual suspects at breakfast are small biscotti-like cookies, called koulourakia, with nuts and raisins or flavoured with cinnamon, cloves, grape must, or orange rind.

Making your own yoghurt, it is very easy

In Greece, and in most of the Greek islands, yoghurt is usually made from sheep's milk. It has a strong flavour and it is healthy, rich in nutrients and thick in texture. We Greeks eat great quantities of yoghurt, plain or mixed with honey and nuts, or with glyko (syrupy sweet preserves, also known as 'spoon sweets' because you eat them straight from the spoon), or as a savoury dish, mixed with herbs and vegetables, as in tzatziki.

Makes 500ml

*1 litre full-fat milk, preferably
    sheep's or goats'*
*2 tablespoons plain yoghurt
    or from a previous batch
    of homemade yoghurt*

Bring the milk to just under boiling point and then pour it into a large, deep mixing bowl. Let the milk cool to about 42°C. Remove any skin that has formed on the surface.

In another mixing bowl, whisk 2 tablespoons of plain yoghurt at room temperature with a few tablespoons of the milk, then pour this into the rest of the milk and beat again.

Cover with a cloth, place in a warm draught-free place and leave for at least 12 hours or overnight (do not disturb it until the yoghurt thickens).

To make really thick Greek yoghurt, remove the skin from the surface of the yoghurt and pour the yoghurt into a muslin bag. Hang the bag over a bowl and let it drain until you get your desired thickness. You can store this in the fridge for 4–5 days.

Warm bread with feta cheese and peach-watermelon preserve

Far away from the glamour of the better-known Greek islands, Lesvos (or Lesbos) is a collection of fine beaches and hillside towns – quiet, unhurried and far from today's fashion awareness. Mitilini port is a glimpse of Greece as it was 50 years ago. Its main port is filled with tavernas and cafés that invite the locals for a quick shot of coffee and a simple breakfast to spice up the day.

On one occasion, Freddie, my godson, went straight into a taverna kitchen and brought a jar of this preserve to our table. Obviously it was something sweet and he greedily took a whole tablespoonful. He ate it with a proclaimed expression of enjoying something excellent and suddenly said 'Ooops!! That stuff is hot and sweet at the same time!!' He then came right back with his spoon and got another large mouthful.

Makes about 1kg

1.5kg ripe peaches, peeled and
    quartered
300g caster sugar
100g finely chopped almonds
50g honey
pared rind of 1 orange
2 mild red chillies, deseeded
    (optional)
300ml fresh watermelon juice

to serve

slices of bread, preferably warm
slices of feta cheese

Combine the peaches, sugar, chopped almonds and honey in a heavy-based casserole, and stir well. Cover and allow to stand for 45 minutes.

Chop the orange rind and chillies very finely. Place these and the watermelon juice in a medium pan. Bring to the boil, cover, reduce the heat and simmer for 10 minutes, or until the orange rind is tender.

Bring the peach mixture to the boil over medium heat, stirring until the sugar dissolves. Increase the heat to medium-high and cook, uncovered, for 20 minutes, stirring often.

Add the orange mixture and keep on cooking, uncovered, for 25–30 minutes, stirring and skimming off the foam with a wooden spoon.

Quickly pour the hot mixture into hot sterilized jars, leaving very little headspace, and wipe the jar rims. Cover and seal at once with lids. Process the jars in boiling water for 5–10 minutes.

To serve, smear the preserve on the bread and add a slice of feta cheese.

Greek yoghurt with roasted pistachio biscuits and cherry butter spread

Serves 4

*500ml Greek yoghurt, bought
   or made as described on page 10*

**For the cherry butter
(makes about 1.5kg)**

2kg seasonal cherries, pitted
500g caster sugar
2 tablespoons lemon juice

**For the pistachio biscuits
(makes about 30)**

50g caster sugar
50g icing sugar
*160g unsalted butter, plus more for
   the baking sheets*
*1 vanilla pod or 1 teaspoon of good
   vanilla essence*
1 egg
200g self-raising flour
*75g coarsely chopped roasted
   pistachios*
50g aniseed seeds
sprinkling of ground cinnamon

First make the cherry butter: combine the cherries with other ingredients in a large heavy-based saucepan. Bring to a simmer over medium heat, stirring often, and cook for about 1 hour or until thickened. A spoonful on a chilled saucer should remain smooth with no signs of wateriness. Spoon into hot sterilized jars to within 1cm from the top and seal.

To make the pistachio biscuits: put both types of sugar and the butter in a large mixing bowl. Using an electric hand blender, cream together well. Mix in the vanilla, egg and finally the flour. Because is a biscuit dough, it softens easily, therefore put in the fridge for 30 minutes or so to make it easier to handle.

Preheat the oven to 180°C/gas 4. Grease 2 or 3 baking sheets with butter and line with baking parchment.

If you have a biscuit set, use it, otherwise take little knobs of dough and roll them with your hand into small walnut-sized balls. Flatten them to form circles about 1cm thick. Line them up on your worktop and press pistachios into one side and aniseed seeds into the other. Place them, pistachio side down, on the prepared baking sheets, leaving about 5cm in between them for expansion during cooking, and sprinkle the tops with a very little cinnamon.

Bake for about 10 minutes or until they turn a good brown colour. Transfer the biscuits to a wire rack and let them cool down.

Fill a bowl with the yoghurt, add a couple of tablespoons of the cherry butter, or more if you prefer it sweet, and then crush over the top 2 or 3 of the biscuits.

PS You can use any other fruits for the butter, and adjust sugar accordingly.

Every June, the village of Karanos on the island of Crete celebrates its cherry harvest. The celebration begins with the obvious blessing, a cherry eating contest and then the locals welcome about a thousand guests to a banquet in which many local delicacies are served. Entertainment in the form of a poetry competition(!), solo singing and non-stop dancing make the party last until sunrise.

Aegina pistachios are famous worldwide as one of the best varieties of the nut. Considered 'must-haves' by gourmets, they come from the island of Aegina, just an hour sailing from Athens. On a good day with a good visibility, if you stand on the Acropolis and look towards the sea, you will notice an island a few miles off the coast. This is Aegina, in ancient times a city-state in its own right and at times a real thorn in the side of Athens. When you get off the boat in Aegina and take a walk around, you will quickly be made aware that you have arrived in the kingdom of pistachios!

Whatever its origins, trahana appears in cookbooks from Hungary to the Middle East and there are dozens of different preparations. It has always been the food of the shepherds, who needed something quick and substantial to prepare that was also easy to carry between their low land in the winter, and high-land homes in the summer. On the Islands, trahana goes into myriad pies in place of rice, especially when greens are used raw.

The people of Crete have a more substantial breakfast, which is a steaming plate of trahana porridge with a cheese spread made out of feta and yoghurt. Often, they also drink a glass of tsikoudia (raki, an eau-de-vie or grappa), especially on colder mornings at sea. Like wine, it soothes the stomach!

Trahana soup with manouri cheese, thyme honey and strawberries

The celebrated British food historian, Elizabeth Luard, calls trahana 'the most primitive noodle in the world, the ancient solution to the problem of how to make milled grain palatable, storable, and portable'. Trahana is a pasta native to Greece and other surrounding areas. Made from flour or bulgur wheat, it has a grainy appearance rather like that of couscous.

There are two basic types, sweet and sour. The sour variety is mixed with thick goats'-milk yoghurt before it is dried, giving it a distinctive tangy flavour. The sweet version is not actually that sweet, but certainly not sour, so is used in a wide range of dishes, both sweet and savoury.

Serves 6

700ml full-fat milk
200g sour trahana
200g manouri cheese (if unavailable, use mascarpone)
250g thyme honey (or more if you like it sweeter)
400g strawberries, chopped

Bring the milk and 700ml water to the boil in a medium heavy-based saucepan. Reduce the heat, add the trahana and cheese, and simmer, stirring occasionally, until the trahana is tender and the mix is thick like porridge. Remove from the heat and serve in individual bowls.

Drizzle in a tablespoon of honey and serve with couple of spoonfuls of chopped strawberries or any other seasonal fruits.

Fried eggs with chopped throumbes olives and tomatoes

Throumbes are not actually a variety of olive, but olives that are left on the trees for much longer than is usual, until they wither and eventually fall. Their very dark black and wrinkled appearance is hardly appetizing, but their flavour is very intense and sweet. The islands of Thassos and Samos produce the best-quality throumbes.

Serves 2-4
20g cold butter
1 tablespoon olive oil
4 eggs
4 slices of bread
4 tablespoons yoghurt

For the coarse olive pâté
250g throumbes olives (if
  unavailable, use Kalamata
  olives)
50g chopped fresh oregano
3 tablespoons olive oil (preferably
  extra virgin)
1 tablespoon breadcrumbs
Juice and grated zest of 1 lemon
15g butter
Coarsely ground black pepper

For the chopped tomatoes
300g finely chopped tomatoes
2 spring onions, finely chopped
50g breakfast radishes, finely chopped
100g anthotiro cheese (if
  unavailable, use ricotta)
60g finely chopped fresh basil
100ml extra virgin olive oil
Salt and black pepper

First make the coarse olive pâté: pit the olives and chop them very finely so as to make a fine mixture. Add the remaining ingredients and mix thoroughly. The pâté can be used immediately, or stored in a tightly covered container in the fridge for several weeks.

Then make the chopped tomatoes, toss all the ingredients together in a bowl, add seasoning to taste and set aside. (If you want to keep the mixture for more than an hour, do not add any seasoning.)

Preferably in a heavy-based frying pan, melt the cold butter with the olive oil and gently fry the eggs over medium heat, until the whites are set but the yolk is still runny.

You can serve all the elements side by side so people can help themselves, or alternatively spread each slice of bread with a couple of tablespoons of the chopped tomatoes, one tablespoon of yoghurt and then place on each slice one of the fried eggs. On top of the egg, add a teaspoon of the chopped olive pâté. Serve immediately, while still warm.

Greek coffee

To prepare Greek coffee you need a narrow-topped small boiling pot called a *briki*, a teaspoon and heat! The coffee is served in cups similar in size to Italian espresso cups. Choose a pot to be close to the total volume of the cups you will prepare, since too large a pot will encourage the precious foam to stick to the inside of it.

A well-prepared Greek coffee has a thick cream at the top, is drunk slowly and is usually served with a glass of cold water All the coffee in the pot is poured into cups, but not all of it is drunk. The thick layer of sludgy grounds at the bottom of the cup is left behind. The cup is then commonly turned over into the saucer to cool, and then the patterns of the coffee grounds can be used for a kind of fortune telling called tasseomancy. The drinker of the coffee cannot read his or her own cup!

Makes 2 cups
4 heaped teaspoons Greek coffee
2 teaspoons sugar (for a medium-
*   sweet coffee)*

The best Greek coffee is made from freshly roasted beans that are ground just before brewing. For best results, the water must first be cold and the coffee and sugar added to the water, rather than being put into the pot first. There are four degrees of sweetness: *sketos* (no sugar), *me oligi* (little sugar), *metrios* (medium sugar) and *vari glykos* (a lot of sugar). The coffee and the desired amount of sugar are stirred until all coffee sinks and the sugar is dissolved. No stirring is done beyond this point, as it would dissolve the foam.

Just as the coffee begins boiling, the pot is removed from the heat and the coffee is poured into the cups.

Warm mastic, honey and peach avgolemono drink

Mastic is a small evergreen tree that is mainly cultivated for its aromatic resin on the Greek island of Chios. The resin bleeds from small cuts made in the bark and is then collected. When chewed, the resin becomes bright white and opaque.

Having Freddie on our trip to Chios was so easy; he was adopted several times by someone from the next table. He sat in their lap and they entertained each other. His best time was when Greek women would pick him up and dance with him around the room, swaying with exaggerated movement until they were both laughing. I tried to think of a similar situation back in London! Everyone might have felt nervous about such a scene.

Serves 4–6
500ml full-fat milk
100g fir pine honey
15g mastic crystals, wrapped
*   in muslin*
4 eggs
150g peaches, peeled, stoned
*   and grated*
2 teaspoons lemon juice

Put the milk and honey in a medium saucepan, add the bag of mastic and place over low to medium heat until honey dissolves, about 5–7 minutes. Remove the saucepan from the heat and set it aside.

In a mixing bowl, lightly whisk the eggs, then pour about 200ml of the milk into the eggs and keep whisking for a couple of minutes. Return the egg-and-milk mixture to the saucepan and cook over medium-to-low heat, stirring gently and continuously until thick enough to coat the spoon about to turn into custard, about 25 minutes.

Add the grated peaches, mix well, remove the mastic bag and pour the fruity custard into a serving bowl, let it cool down for 15 minutes and serve. Freddie likes it with a drop of lemon juice.

Bougatsa (warm savoury filo pie)

The sun was about to rise and the street vendors and bakeries were opening their stands for the early morning ritual. We chose a café facing the old town of Chania in Crete and, with the coffees we ordered, we started what is an essential part of the early morning, socializing, with seductive flavours and smells savoured through our lowered eyelashes. I looked around and it seemed that for a moment we were all united in one flavour!

Makes 12

2 egg yolks and an extra ½ egg
   white, both at room temperature
60g caster sugar
250g sweet mizithra cheese (if
   unavailable, use mascarpone)
250g anthotiro cheese (if
   unavailable, use ricotta)
100g feta cheese
½ teaspoon ground cinnamon, plus
   more for serving if you wish
½ nutmeg, freshly grated
Grated zest of 1 lemon
Grated zest of 1 orange
12 sheets of filo pastry
125g butter, melted

Beat the egg yolks, egg white and sugar until thick, creamy and pure white; this will take a few minutes.

Using an electric mixer, beat the sweet mizithra at high speed until light and fluffy. Lower the speed and add the anthotiro, feta, cinnamon, nutmeg and citrus zests. Beat for a couple of minutes at high speed, and then set aside while you are preparing the filo.

Preheat the oven to 180°C/gas 4.

Unroll one sheet of filo on a worktop, with the narrow end facing you. Keep the remaining sheets covered with a damp kitchen towel. Brush the sheet lightly with melted butter. Place about 3 tablespoons of the cheese mixture on the lower third of the filo and spread evenly, nearly covering the lower third of the pastry. Fold the right and left sides of the filo in towards the centre so that the edges just meet. Lightly brush the folded sides with butter. Fold the lower third up and brush with butter. Fold the upper third down and brush with butter. It should now look like an envelope. Place on a lightly buttered baking sheet and repeat the procedure until you have used all the pastry and all the cheese mixture.

Brush the tops of the bougatsa with any remaining butter and bake until golden-brown, about 15-18 minutes. Allow to cool for 20 minutes before serving warm. If you like them a bit sweeter, dust them with icing sugar and some more cinnamon.

To Make Tea 'The Greek Way'

With hills and crevices of hard limestone full of a mad variety of wild plants, Greeks have always had a plethora of seeds, roots and flowers from which to make different kinds of teas. The common ones use sage, lemon blossom, mint, camomile, poppy and, I believe, lots more that I am not aware of.

Makes 4 cups

15g dried leaves and flowers
750ml boiling water

I still get so excited with this most magical scent that wafts up as I pour on piping hot water; a sort of woodsy/lemony smell.

Strain and add a little lemon juice, honey or sugar, or cinnamon or drink just as it is. I drink mine with throumbes olives, sweet mizithra cheese (like mascarpone) and crusty bread.

Omelette with honey and sesame seeds

The Greek islands have been the crossroads of the Mediterranean since the time of Homer. Over the centuries, Phoenicians, Athenians, Macedonians, Romans, Byzantines, Venetians, Ottoman Turks and Italians all stirred the pot in this region, putting their distinctive stamp on the food, and this recipe is a good example of such a heritage.

Serves 6-8

10 eggs
50ml milk
100ml olive oil
50g butter
200g tomatoes, skinned, deseeded
    and coarsely chopped
1/2 bunch of mint, picked and finely
    shredded
Salt and black pepper
200g feta cheese
Generous sprinkling of sesame seeds
150g thyme honey
Juice of 1 lemon

Beat the eggs together with milk in a medium-sized bowl until well mixed but not foamy.

Heat the olive oil and the butter in a large heavy-based frying pan over medium heat. Add the chopped tomatoes and mint, season and stir for 3-4 minutes. Add the feta, pour the egg mixture into the pan and cook for a few minutes, pushing the edges towards the centre with a spatula but without disturbing the bottom. Cook for about 4-5 minutes more, until the top is no longer runny and the bottom is fairly well coloured. Sprinkle over the sesame seeds, then drizzle the honey over the top and finally fold over to make a half-moon shape.

Drizzle the lemon juice on top and serve right away, with warm bread and sweet-cured black olives.

2

Spring Tides, Full Moon and Seasonal Vegetables

The islands of the 'agoni grammi' (meaning 'off-the-beaten-track line') are all the small islands that would otherwise be neglected by the ferry lines but for the Greek government forcing them to make quick visits to them. Usually the oldest boats are allocated to these routes.

If you could hover, god-like, high above these islands with a telescope, around the last week of February you would observe that life was still at a winter's snail's pace. As the weeks move towards Easter's full moon, however, the pace quickens. The islands awake, revealing beautifully the dressed-up gardens, the local markets filling up with the local produce and the baker preparing his ovens for the racing track of the coming carnival in the three weeks before Lent. As the spring's first vegetables arrive, the islands begin to resemble soup plates full of life.

Cretan Dolmades
Young vine leaves stuffed with bulgur wheat and trahana

Mixing flour, yoghurt or sour milk, salt and spices produces what we call trahana (see pages 14–15). The mixture is then allowed to ferment, then dried, ground and sieved. The fermentation produces lactic acid and other compounds which give the trahana its characteristic taste. As trahana is both acid and low in moisture, it preserves its milk proteins well for long periods.

Young vine leaves preserved in a light brine should be a very pale green in colour and if they are not then remove the leaves from the jar and rinse them very well under cold water.

Serves 8 (makes about 64)

500g preserved young vine leaves
250ml extra virgin olive oil
200g onions, finely chopped
100g spring onions, finely chopped
(including most of the green)
200g bulgur wheat
50g trahana (if unavailable, use
more bulgur)
100g currants
100g pine nuts
100ml lemon juice
Salt and freshly ground black pepper
300ml hot water
100g fresh mint, finely chopped

Bring a pot of water to the boil and blanch the vine leaves, in batches, for about 5 minutes each batch. Place the blanched leaves in a colander and rinse with cold water.

Heat 100ml olive oil in a large heavy-based casserole and add the onions and spring onions. Toss to coat with the oil and sauté over low heat for about 10 minutes, until translucent. Keep everything moving as you add the bulgur wheat, trahana, currants, pine nuts, lemon juice, 100ml more olive oil, salt and pepper. Finally, add the hot water, cover and simmer for 5 minutes. Remove from the heat and allow to cool. When cool, add the chopped mint and mix well.

Separate any ripped or misshapen leaves from the rest. Pour the remaining olive oil on the bottom of a large heavy-based casserole and tilt around so that the oil spreads over the whole surface. Spread a few of the ripped or misshapen leaves on top, just enough to cover the surface.

Arrange the other whole leaves, veiny side up, as many leaves as will fit spread out on the kitchen table. Snip off the tough stems if there are any and place about 1 teaspoon of filling on the centre of each leaf. Fold in the sides and roll up from the base, tucking in the sides a little as you go.

Place the rolled stuffed leaves snugly next to one another, seam-side down, in the pan, in several layers if that proves necessary. Add just enough water to cover. Place a plate inside the pot over the vine leaves as a weight to keep them from opening up during cooking. Cover the pot with its lid and simmer the dolmades over low heat for about 40 minutes, until the leaves are tender and the filling cooked.

Remove, allow to cool slightly and serve.

Santorinian oven-roasted tomatoes and peppers with cinnamon

The island of Santorini is more renowned for its dramatic setting, volcanic flows and scatterings of ash than it is for its agricultural products, most of which seem to be associated one way or another with the volcanic soil. One of these is the diminutive 'waterless' tomato. The Santorinian tomato actually belongs to a different species and comes in two varieties. The 'authentic type', in which the rounded sides of the tomato are fluted vertically, giving the impression of a segmented fruit, and the 'kotiko' type that does not have any such flutings. The sun-dried tomatoes you need for this dish are those lovely plump, moist ones that need no rehydration.

Serves 4–6

*250g plump, moist ready-to-eat
    sun-dried tomatoes*
300g preserved red peppers
200ml extra virgin olive oil
¹/₂ teaspoon ground cinnamon
*Salt and freshly ground black
    pepper*

**For the oven-roasted
tomatoes**

2 tablespoons olive oil
500g tomatoes
1 tablespoon sea salt
¹/₂ tablespoon dried thyme

First roast the tomatoes. Preheat the oven to 80°C/gas ¹/₄. Brush a roasting tray lightly with olive oil. Cut the tomatoes across in half and place the halves very close to one another in a single layer on the roasting tray, cut sides up. Sprinkle the salt, thyme and remaining olive oil over them.

Place in the oven and leave for about 4 hours, but check after 3. The time can vary slightly according to the quality of the tomatoes and the oven. The tomatoes should not dry out completely but should be chewy. If you are in a hurry, buy more good ready-to-eat sun-dried tomatoes instead, so they won't need roasting.

Purée 200–250g of the roasted tomatoes in a blender with the sun-dried tomatoes until smooth; gradually add the oil, cinnamon, salt and pepper.

The purée may now be put in a sealed container and stored in the fridge, where it will keep for a week. It can also be covered with olive oil before being sealed and refrigerated, and this way it will last about 3–4 weeks.

Use instead of ketchup or stir into pasta, or simply serve with warm bread and feta cheese.

The Santorinian tomato bears more fruits than ordinary tomato plants, matures earlier in the year, maintains its deep red colour and requires no watering, an important attribute for the Aegean islands, where water is often in short supply. Needless to say, it tastes like a real tomato!

Bulgur, walnut and spinach pilaf with aged red wine roast tomatoes

I cheated on a few of the steps here because the island we were visiting when I made it was suffering from a long power cut, so we couldn't use one of the local kitchens. Everyone was, however, in a holiday mood. I had just one pot and one camping fire so I made these in advance.

In Crete, bulgur – or *pourgouri* as it is known in Greece – is widely used and traditionally the grain is parboiled, then dried, usually by spreading it in the sun, and then de-branned. Bulgur is often confused with cracked wheat, which is made from crushed wheat grains, which have not been parboiled.

Serves 6

300g onions, finely chopped
120ml olive oil
3 garlic cloves, crushed
200g bulgur
300ml hot chicken stock
50g butter, cut into small pieces and chilled
70g walnuts, crushed
1/2 teaspoon freshly ground cinnamon
70g mint, finely chopped
250g spinach, finely shredded
Juice of 1 lemon

For the roast tomatoes

1kg tomatoes
120ml olive oil
4 tablespoons aged red wine vinegar or balsamic vinegar
10g mountain thyme
Coarse sea salt and black pepper
2 tablespoons petimezi (naturally sweet grape juice; if unavailable, use 2 teaspoons soft brown sugar)

First roast the tomatoes: preheat the oven to 160°C/gas 3. Halve the tomatoes lengthwise and put them in a small roasting pan. Mix together the olive oil, aged red wine vinegar, thyme, salt and pepper. Pour this over the tomatoes. Turn them over, making sure they get well coated in the mixture, ending up with them cut-side up. Drizzle the petimezi over the top and put the tray in the preheated oven. Cook for 45 minutes until the tomatoes are shrunken and sweet.

To make the pilaf: in a heavy-based casserole, sauté the chopped onions in the olive oil. When these are soft and translucent, add the garlic and cook for another couple of minutes. Tip the bulgur wheat into the casserole; pour in the stock and season. Bring to the boil, then turn down the heat and let the bulgur simmer in the stock for about 15 minutes. All the stock will have been absorbed by then. Dot the pieces of butter pieces all over the surface of the bulgur, sprinkle over the crushed walnuts and cinnamon, cover the pot and let the bulgur sit to fluff up for another 10–15 minutes.

Fluff the bulgur wheat with a fork, place in a shallow bowl, and sprinkle on half of the mint, followed by the shredded spinach and the tomatoes (they can be hot or at room temperature), then finally the rest of the mint. Drizzle over some lemon juice and serve.

Samothraki, with its hulky and sulky air, is in the northern Aegean and is one of the destinations that I would rather like to keep secret. It is a paradise for nature lovers who feel the need for peace and quiet, or people who delight in drinking clean cool stream water. The water from the mains supply is also clean, as it comes from the same spring!

The sweetbreads served on the islands are usually from lambs or goats and need soaking in water before they're peeled, a process that transforms the chewy membrane into a tender delicacy. This local creation requires two pans; two different recipes and several nimble fingers, but the result will produce a many-voiced full-volume conversation. Use good-quality frozen broad beans as the freezing tenderizes their skin.

Broad beans with braised lamb sweetbreads, yoghurt, parsley and dill

Serves 6-8

500g lamb sweetbreads
100ml olive oil
250g spring onions, both white and green parts, thinly sliced
300g tomatoes, deseeded and grated
1kg frozen broad beans
70g dill, finely chopped
70g flat-leaf parsley, finely chopped
Salt and freshly ground pepper
100g butter
A little plain flour
200g leeks, both white and green parts, finely chopped
20g marjoram leaves
150ml chicken stock
50ml Muscat wine or sweet sherry
100g yoghurt

Place the sweetbreads in a large bowl, add cold water to cover and set aside for an hour. Rinse them and repeat the same process a couple more times or until the water remains fairly clear and the sweetbreads look blanched and not pink. Peel off the outer membrane that covers the sweetbreads.

In a large heavy-based casserole, heat the olive oil over medium heat. Stir in the spring onion and sauté until wilted, about 5 minutes. Stir in the tomatoes with their juices and some seasoning. Bring to the boil, then reduce the heat and simmer until the sauce has thickened, about 15 minutes. Add 500ml water and the broad beans, and simmer for 45 minutes. Add the chopped herbs, adjust the seasoning, mix well and set aside while you prepare the sweetbreads.

Melt the butter in a large heavy-based frying pan over medium heat. Pat the sweetbreads dry, roll them in the flour and sauté until slightly coloured, about 3 minutes. Add the leeks, marjoram, chicken stock, Muscat and seasoning. Give everything another good stir, cover the pan and simmer gently over a low heat for 15 minutes.

Lift the lid and, with a ladle, remove some of the juices into a mixing bowl. Add the yoghurt and whisk well. Pour the yoghurt sauce back into the frying pan, remove the pan from the heat and set it aside for 10 minutes, so the sauce can mellow.

When ready to serve, transfer the broad beans to a serving platter, spoon the sweetbreads on the top and moisten them with some of the yoghurty juices.

Okra and chicken filo rolls

Okra must be picked young. After the pod is more than 4 days old it becomes almost useless for cooking. If you buy fresh okra, try to get young pods without any signs of bruising. They should be tender but not soft and no longer than 5cm, as greater length is a sign that they have been left too long before picking. Serving okra often divides the dinner table into two seemingly irreconcilable camps: fervent fans who adore its slightly viscous texture and ardent detractors who don't! Greeks generally adore okra.

Serves 5-6

400g very small okra
200ml white wine vinegar
500g chicken legs
Salt and coarsely ground black
    pepper
5g mountain thyme
Juice of 1 lemon
150g red onions, peeled and grated
100g peppers, deseeded and finely
    diced
150ml olive oil
300g tomatoes, deseeded and grated
70g flat-leaf parsley, finely chopped
500g filo pastry
2 egg yolks
2 tablespoons milk
100g butter, melted
100g sesame seeds

Okra is fairly fragile and breaks easily during cooking, so use a very sharp paring knife to peel away the wider ring at the top of the pod as well as the cone tip, without piercing the pod itself. When you do this, none of the interior gluey substance gets released. Lastly, toss with the vinegar, let stand for at least half an hour, then rinse and towel dry thoroughly before use.

Preheat the oven to 180°C/gas 4.

In a mixing bowl, toss the chicken legs with some seasoning, the thyme and lemon juice. Set aside.

In a large pan over low heat, gently fry the grated onions and peppers in the olive oil, stirring them constantly to colour them very lightly all over. Add the grated tomatoes, cook for 5 minutes and then add the chicken. Mix well and remove from the heat. Arrange the okra in a single layer in a roasting tray. Place the chicken legs on the top of the okra and then pour over the tomato sauce. Bake for about 40 minutes. If by this time the sauce is still watery (which will make the filo soggy), remove the chicken and continue to cook until the sauce has thickened up. Return the chicken to the tray, add the chopped parsley and allow to cool down. Leave the oven on.

When the chicken is cold enough to handle, pick off the flesh and discard the bones, etc. Mix into the okra and tomato sauce gently so as not to break the okra.

Open out the filo sheets, leave them in a pile and cover with a damp cloth so they don't dry out. Mix the egg yolks with the milk in small bowl. Place one sheet of filo on a work surface, brush it with melted butter and then the egg wash and lay another sheet on top.

Divide the okra and chicken mixture into six equal portions. Spread one portion of the mixture along the edge of the long edge of the filo. Roll the sheet up from that end to enclose the filling, tucking in the edges as you go, and brushing thoroughly with butter and egg wash as you go. Repeat with remaining sheets of filo and filling to make six long rolls in total.

Place the rolls side by side on a greased baking tray, brush the tops with melted butter and egg wash, sprinkle over some sesame seeds and bake in the preheated oven for about 30 minutes or until golden.

Serve warm with the Beetroot with a manouri and yoghurt dressing on pages 44–5.

Beetroot with a manouri and yoghurt dressing and fried marida (silver-headed Aegean whitebait)

Manouri is a semi-soft fresh white cheese made from the drained feta whey. It has a slightly grainy, creamy texture and a milky taste, with a mild citrus hint. Substitutes for manouri can be a good cream cheese or chèvre.

Serves 6

500g beetroots
200g manouri cheese, crumbled
150g Greek yogurt
100g spring onions, thinly sliced
*½ teaspoon cumin seeds, tossed
    and ground*
Sea salt and coarse black pepper

For the marida

*About 500ml olive oil or good-
    quality vegetable oil for frying*
500g silver-headed whitebait
200g flour
100g fine semolina
Sea salt and black pepper
6 lemon wedges

Start by preparing the beetroot: wash them well and trim the stems. The quick way to prepare them is to boil them, but if you have enough time roast them. Cover them with a couple of layers of kitchen foil, place the parcel on a roasting tray and roast in an oven preheated to 200°C/gas 6 for a couple of hours or until they feel tender when cut with a knife. Remove them from the oven and let them cool down. Boiled or roasted, when cold enough to handle, peel them and grate the beetroot coarsely into a mixing bowl.

Put the manouri and yoghurt in a blender and whiz until you have a smooth sauce. Pour the manouri sauce into the mixing bowl, add the chopped spring onions, cumin and some seasoning. Mix well.

Prepare the marida: heat the oil in a heavy pan that gives it a depth of about 3cm. Roll the fish in a mixture of the flour and semolina so that they are well coated, then shake off any excess. Drop them gently into the hot oil and move them around while they fry. Fish them out once they are slightly browned and crisp. Season them while still hot.

Serve with lemon wedges and the beetroot salad.

Aubergine Stifado

Serves 6

*300–400ml olive oil or good-
  quality vegetable oil*
300g button onions, peeled
*Salt and freshly ground black
  pepper*
1kg aubergines
100ml red wine
*300g tomatoes, skinned, deseeded
  and grated*
*1 teaspoon cumin seeds, briefly
  roasted in a dry frying pan and
  finely ground*
1 cinnamon stick, finely ground
*2 tablespoons oregano (preferably
  Greek mountain)*
*4 teaspoons aged red wine vinegar
  or balsamic vinegar*
150g basil leaves, shredded

Preheat the oven to 180°C/gas 4.

Put 70ml of the olive oil in a heavy-based baking tray, add the
button onions with a very little seasoning, and roast them for about
30–45 minutes.

Peel half the skin off the aubergines in long strips so that they look
stripy. Cut them into 2cm dice, sprinkle these with salt and allow
them to stand, draining in a colander, for about 30 minutes.

Pat the diced aubergines dry and fry in small batches using the
remaining olive oil. Do not let the oil smoke at any point. Remove
and drain on kitchen paper. Discard the oil and deglaze the pan(s)
with the red wine. Set this reduction aside.

Remove the tray of onions from the oven, leaving the oven on.
Add the aubergines, grated tomatoes, spices, red wine reduction
and oregano to the onions and mix thoroughly, so the liquid is
distributed equally within the vegetables. Cover the tray with
baking parchment and kitchen foil, and place back in the oven.
Bake for 1½–2 hours in the oven on the same heat setting. Remove
from the oven, add the vinegar and shredded basil and allow to
rest for at least 15–20 minutes before serving.

Cretan snails with stoneground wheat

After exploring and eating around the island, we were reminded again why we Greeks consider Crete our culinary castle, where the food tradition of the whole Aegean is gathered. Crete's well-defined cuisine is ruled absolutely by the seasons and a native versatility. Although the same simple ingredients are used again and again, local kitchens use them to create a variety of memorable dishes.

Serves 5–6

60 large snails
125g coarse sea salt
200g onions, peeled and grated
6 garlic cloves, crushed
200ml olive oil
750g ripe tomatoes, skinned,
*   deseeded and grated*
250g coarse stoneground wheat (if
*   unavailable, use bulgur wheat)*
100g flat-leaf parsley, finely chopped
100g rocket, finely chopped
Freshly ground black pepper

Rinse the snails several times with cold water. If any snails no longer have their lip covering, or if they have withdrawn into the shell, they must be discarded as they are dead.

Place a large casserole filled with water over medium heat. When warm and well before it boils, add the snails and about 100g salt. When the water starts to boil, reduce the heat and simmer for 15 minutes, uncovered. Drain in a colander under cold running water until water runs clean. With a sharp knife, remove the lips and let the snails drain for 15 minutes.

In a heavy-based casserole, sauté the onions, garlic and snails in the olive oil over medium heat, stirring frequently, for 5 minutes. Stir in the tomatoes, reduce the heat to low, cover and simmer for 10 minutes. With a slotted spoon, remove the snails and set aside in a covered dish.

Pour 1 litre of water into the casserole and when it is about to come to boil add the wheat and 25g salt. Stir constantly for the first few minutes to prevent clumping. Cover and continue to cook over low heat for about 15 minutes, stirring frequently. Remove from the heat and adjust the seasoning, then stir in the snails, the chopped parsley, rocket and pepper to taste.

Cover and allow to rest for 20 minutes before serving. The pilaf should be slightly soupy.

Symi shrimps with tomatoes and caper leaves

Serves 4–6

250ml olive oil

300g button onions, peeled and left whole

15 garlic cloves, 10 left whole (but peeled) and 5 crushed

750g ripe tomatoes, deseeded and coarsely grated

150g preserved caper leaves (if unavailable, use small capers)

Salt and coarsely ground black pepper

500g unpeeled whole raw prawns

100g rusks, coarsely crumbled (if unavailable, use good breadcrumbs)

Heat 150ml olive oil in a large heavy-based frying pan and cook the whole onions and whole garlic cloves over a very low heat for 20 minutes. Add the tomatoes and cook for 30–45 minutes until the liquid begins to thicken. Stir in the caper leaves, take the pan off the heat and add some coarse black pepper and little salt (as the caper leaves are preserved, they are already salty).

Heat another large heavy-based frying pan, add one-third of the crushed garlic and the remaining olive oil and cook for 2 moments only (10–15 seconds) over a medium heat, without allowing the garlic to colour.

Season the prawns and put as many as you can get in a single layer into the pan. When they are seared and turned dark pink (this will take couple of minutes), press the head and shell to release some of their juices. You do this to keep your pan clean. Set the cooked prawns aside in a warm place, wipe the frying pan with some kitchen paper and repeat until all prawns are cooked.

Serve the prawns on top of the tomato and caper leaf sauce, with the crushed rusks sprinkled on top.

Borlotti beans with grilled chicken wings and purslane

Serves 6

24 chicken wings

For the marinade

4 tablespoons olive oil
Juice of 1 lemon
50g flat-leaf parsley, finely chopped
*Salt and freshly ground black
    pepper*

For the beans

*500g fresh borlotti beans (300g
    when podded)*
*250g button onions, peeled and
    left whole*
100g carrots, peeled and finely diced
*2 celery stalks, trimmed and finely
    diced*
6 sage leaves
*200ml olive oil, plus more for
    drizzling*
4 garlic cloves, crushed
250g purslane, trimmed
Juice of 1 lemon

Make the marinade by mixing the olive oil, lemon juice, parsley and seasoning, and place the chicken wings in the marinade. Leave them, covered, in the fridge until the borlotti beans are ready.

To prepare the beans, put them in a large casserole with the button onions, carrots, celery, sage and 50ml olive oil. Do not add any salt until the beans are almost cooked, otherwise their exteriors will stay hard. Add twice the beans' volume of water, cover and bring to boil. When it comes to boil, remove the lid, skim off the foam, lower the heat and simmer for just over an hour. Drain off most of the liquid from the beans but leaving them nice and moist, reserving the cooking liquid.

Preheat a medium grill.

In the same casserole, heat the garlic in the remaining olive oil for 10–15 seconds only. Add the purslane, put the lid back and cook over very low heat for a minute without adding any water. Stir in the beans and lemon juice, mix well and adjust the seasoning.

Place the chicken wings on the top of the cooked beans, pour the marinade over them and cook under the grill for 6-8 minutes, turning them only once.

Serve the beans and wings drizzled liberally with extra virgin olive oil. If you need the beans to be saucier, add some of the reserved bean cooking liquid.

Kythira is the holy grail of island hopping! We spent a few days there during the wine festival in Mitata, and then went on to the island of Antikythira, the place where, in the early 1900s, a fisherman found what was thought to be an astrolabe, a device to measure the altitudes of celestial bodies, that dated back to ancient Greek times. When the instrument was found, it had a lot of metal wheels arranged in a way that simulates the movement of the stars and does the required calculations! Nowadays, it is clear that is not, in fact, an astrolabe but a kind of astrological calculator. Who designed it and who made it with such accuracy remains a mystery.

As we arrived on this most remote island in the Ionian Sea, which has one teacher, a handful of pupils, one doctor, one policeman, one telephone and one monastery, we came across a mountain of *barbounia* (borlotti beans), which normally appear in the market during the mid-summer. With no bank to get some money, I ended up in the only kitchen with the time-consuming job of peeling the pods, a job which is normally reserved for pairs of lovers. The islanders' generosity also brought us a bowl of free-range chicken wings. What can I say? The results were truly amazing.

3

Perfidious Greek Herbs

From the gentle whiffs of the sun-drenched mountain oregano and thyme you are able

to guess the entire recipe – roast potatoes with lamb drippings, lemon and thyme.

Remarkable? Not really, that is the all-pervasive aroma of many of the forgotten

Cycladic islands. Throughout recorded history these islands have never been antagonists

with each other. Indeed, quite the opposite; they have always extended their hands

to one another and shared local ingredients and knowledge.

When I look at the map of Cyclades, to me it is a perfect and beautiful puzzle and,

as the sun sets in the centre of all these islands, they are all about to keep changing colours:

moody and dull golden, green, blue and red colours paint the rocks differently.

Then a strange serenity cuddles the islands and at last there is happiness on our faces

and some are too close to tears.

Button onion bulbs with dill

Peel my winter island patience like an onion! There is a certain hypnotic melancholy in the islands through the cold and windy months. All the inhabitants, who manage year in year out to reach the brilliance of the spring with even more grace and beauty, make me think of giant sea birds refusing to relent in the face of incredible obstacles to get to their warm-weather destinations. This recipe is from the unrewarding and ragged island of Ikaria, with its striking cliffs covered with a mythical mist.

For me definitely the easiest way to peel button onions is to place them in a saucepan, just cover with boiling water and leave for a minute or two. Drain the water and then squeeze the silver skin with your fingers. Simple!

Serves 6

1kg button onions, peeled and left whole
100ml red wine vinegar
100ml olive oil
100g dill, finely chopped
Sea salt and coarsely ground black pepper

Place the onions in a casserole and boil in a mixture of a little water and half the vinegar for 20–30 minutes.

Beat together the remaining red wine vinegar, the oil, dill, salt and pepper. Set aside.

Drain the boiled onions, place in a bowl and then pour the olive oil and vinegar dressing over the top of them.

Serve warm or cold with feta cheese, radishes and mixed olives.

Bougiourdi
(roast feta en papillote)

Serves 6

*100g pickled green chilli peppers,
   deseeded and finely diced*
*200g large tomatoes, skinned,
   deseeded and coarsely chopped*
50g spring onions, finely chopped
100ml extra virgin olive oil
*75g basil, leaves picked and stems
   chopped finely*
300g barrel-aged feta

Preheat the oven to 200°C/gas 6.

Put all ingredients except the feta into a mixing bowl and mix well.

Cut the feta into 12 equal slices. Take one slice, spread about one-twelfth of the tomato mix on it, sandwich with another slice of feta, then top that with another twelfth of the mix.

Cut six 20cm squares of baking parchment and 6 similar squares of kitchen foil. Fold one square of parchment in half to create a crease, and then open up. Place one of the feta 'sandwiches' next to the crease. Repeat with the remaining squares and portions of feta.

Working with one square, fold the paper over the feta, and then fold the edges several times, crimping to seal the packet completely. Wrap it again in a piece of foil in the same way. Repeat to make 5 more packets.

Place the packets on a baking sheet and bake for about 10–15 minutes.

Place the packets on serving plates. To serve, slit an X in each packet and fold open.

Kaseropita with basil

Pies remain one of the mainstays of the Greek table, as well as a viable choice of our street food. Fillings change from region to region and our pies contain anything Greece produces, any kind of weird or exotic local delicacies from all corners of the country.

Serves 6

180g butter, plus more for the
    baking dish
500g kataifi dough (shredded filo
    dough)
300g kasseri cheese, grated (if
    unavailable, use mature
    Cheddar)
300g kefalotiri cheese, grated (if
    unavailable, use pecorino)
100g basil
5 eggs
500ml milk
1 teaspoon ground cinnamon

Preheat the oven to 180°C/gas 4 and butter a baking dish no bigger than 30x25cm. Place half of the kataifi dough on the dish. Spread the cheese on top. Pick the basil leaves and arrange them over the cheese, then finely chop the stems and sprinkle over the cheese as well. Cover with the rest of the kataifi dough.

Melt the butter in a small saucepan and spread this over the top of the pie.

Beat the eggs and mix them with the milk and cinnamon. Pour this mixture on top of the pie.

Bake for 30–40 minutes, until the top is lightly coloured and crisp. Remove from the oven and allow to cool down for 30 minutes.

Serve with the mixed green salad from Lesvos (see pages 68–9).

Resembling angel hair pasta or shredded wheat, kataifi is a partially cooked filo-like dough extruded into very fine strands and dried. It is widely used in the food of the Middle East, the Balkans and Turkey, as well as in Greece, usually for the preparation of desserts and sweetmeats.

Lesvos is like an intricate wild botanical garden, with a steep pattern of land that is covered with nettles, wild herbs, spring rock-cress, dropping Star of Bethlehem and the endemic flower of the Aegean Islands, the bee orchid.

Mixed green salad from Lesvos

Serves 6

*1 head of cos lettuce, cored and
leaves separated*
100g rocket, finely chopped
100g watercress, finely chopped
*1 bunch of spring onions, white
and most of the green parts,
thinly sliced*
1 bunch of dill, finely chopped
*1 bunch of mint, tough stems
removed, thinly sliced*
*¹/₂ bunch of borage, coarsely
chopped*
*1 small fennel bulb, very thinly
sliced, preferably with
a mandolin*
120ml extra-virgin olive oil
50ml red wine vinegar
Salt and coarse black pepper
100g toasted pine nuts

Stack half the lettuce leaves, roll them up and cut across into very thin slices. Repeat with the remaining lettuce leaves.

In a large bowl, combine the lettuce and other greens, spring onions, dill, mint, borage and fennel.

In a small bowl, whisk together the oil, vinegar and plenty of black pepper, and pour it over the salad. Add salt to taste. Toss and sprinkle with the pine nuts. Serve at once.

TASTE : 5
EFFORT : 9

DISAPPOINTING, TAKE SEVERAL BIG, TASTY
FLAVOURS, COMBINE & RESULT WAS STRANGELY BLAND ?!

Koliosalata
(smoked mackerel salad dip)

Serves 6 plus
370g smoked mackerel
60ml extra virgin olive oil
Juice of 1 lemon
1 bunch of dill, finely chopped
150g strained Greek yoghurt
Salt and freshly ground black
*   pepper*
Good bread, to serve

Clean the fish; remove its bones and skin.

With a fork, mix the flesh with the oil, lemon juice, dill and yoghurt. Add a little salt (but not too much as the mackerel is already salted) and plenty of black pepper. Mix well in order to make a pulp with a smooth-coarse texture, similar to the texture of the better known taramasalata.

Serve with good bread.

Basil oil

If you can't track down basil oil in your local market, you can make it yourself.

Makes about 500ml
200g basil
30g flat-leaf parsley, picked
400ml extra virgin olive oil

Blanch the basil and parsley in boiling salted water until wilted, no more than 10 seconds. Drain and drop the herbs in ice-cold water.

Drain well again and whiz up with olive oil in a food processor, then strain through a cheese cloth. Skim the clear green oil off the top, discarding the solids, and reserve. The oil will keep for several weeks in a cool dark place.

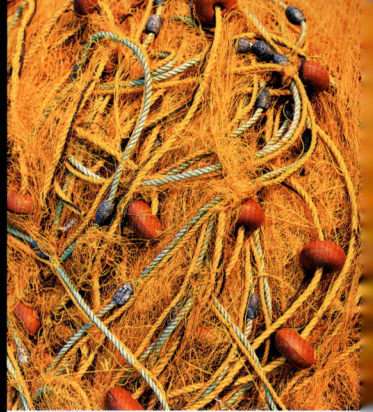

Mackerel fillets with caper leaves and fresh dill

Serves 4

*50g Corinthian sultanas or any
   other good-quality sultanas*
Juice of 2–3 lemons
*1 bunch of spring onions, trimmed
   and finely chopped*
1 bunch of dill, finely chopped
30g caper leaves or capers
50g pine nuts, lightly toasted
150ml extra virgin olive oil
500g skinless mackerel fillet
*Salt and freshly ground black
   pepper*
200g rocket, trimmed and torn

Soak the sultanas in the lemon juice.

Remove any pin bones from the mackerel fillets, trim away any blemishes and cut into slices 5mm thick. Arrange the slices on a serving platter and chill for at least 30 minutes.

In a mixing bowl, put the spring onions, dill, caper leaves, pine nuts, olive oil and the lemon juice with the sultanas. Mix well and coat the mackerel fillets evenly with the mixture. Season to taste and put the platter back in the fridge for an hour or so.

Mix the rocket into the mackerel salad and serve.

Fennel rissoles from Serifos

This is such a simple recipe to introduce the magic of Serifos! In the infinite blue of the Aegean Sea, Serifos – the 'iron island' of the Cyclades – stands out in its unique beauty. It is said to have once been occupied by the Cyclopes. These one-eyed giants, the children of Poseidon, lived in a cave near Psaropyrgos and the huge walls still to be seen all over the island are attributed to them.

For me, what makes this island really magic is the love story of Zeus, who fell in love with Danae. In his desire to hold her in his arms and drink unconditional joy from her lips, he transformed himself into golden rain to seduce her. She later gave birth to Perseus. When her father, King Acrisius, found out, he locked Danae and her son in a box and threw them into the sea. The sea was calmed by the gods and the box and its precious contents were eventually washed up on Serifos.

Serves 4–6

250g fennel, thinly sliced
300–350g flour
1 bunch of spring onions, white and green parts, thinly sliced
½ bunch of tarragon, picked
Salt and freshly ground black pepper
Vegetable oil for frying

Mix well all ingredients except the oil and then pour in as much water as required to make a soft pulp.

Heat the oil in a frying pan and drop heaped spoonfuls of the mixture into the pan. Fry the rissoles until golden brown on both sides, turning once. You may have to cook in several batches.

Serve hot.

Warm salad of potatoes with rosemary and garlic

I had this salad on the island of Folegandros, where at Easter and at certain other times through the year the locals process the icon of the Virgin Mary on a boat. There is a tale that says that when the island was sacked in 1715 by the Ottoman navy, the occupiers removed her gold dress, but by a miracle she returned to the island in her full glory! Since then, once a year and for three days she is taken to visit every one of the freshly white-painted houses on the island, while the locals pay their respect by cooking aromatic honey pies, savoury pies, tsoureki (brioche) and drink tsipouro (grappa).

Often, when I cook potatoes this way, I add a few soft-boiled eggs and a handful of either Santorinian caper leaves or capers.

Serves 6

1kg potatoes
70g garlic, peeled and left whole
300g button onions, peeled and
    left whole
20g rosemary, left whole
150ml olive oil
Juice and grated zest of 2 lemons
Salt and freshly ground black
    pepper
2 bunches of spring onions, finely
    chopped

Preheat the oven to 160°C/gas 3.

Peel the potatoes and cut them into walnut-sized pieces. Place these in a large heavy-based casserole, add the garlic, button onions, rosemary, 150ml water, the olive oil, and the lemon juice and zest with some seasoning.

Cover the casserole well and place in the preheated oven for 2 hours, or until the potatoes are soft.

Remove the casserole from the oven, add the chopped spring onions, adjust the seasoning and mix well, being careful not to break up the potatoes. Allow to cool for 20 minutes with the lid on before serving.

Fragrant nut rice with chicken livers

Livers can be easily overcooked. Generally, the best technique for sautéing them is to cook them very fast in hot oil so that the outside gets sealed and the inside stays moist and a bit pink.

Serves 6

1.5 litres good chicken stock
100ml olive oil
50g butter
50g orzo (rice-shaped pasta; if unavailable, use spaghetti cut into small pieces)
500g medium-grain rice
½ teaspoon ground cinnamon
6 sage leaves
Peel of 1 lemon
100g pine nuts
100g pistachio nuts, shelled
150g flat-leaf parsley, finely chopped
Salt and freshly ground black pepper

For the chicken livers

500g chicken livers
200g plain flour
100ml vegetable oil
50g butter
200g shallots, finely chopped
75ml Mavrodaphni or port
250g tomatoes, skinned, deseeded and chopped to a purée
100ml olive oil

Bring the stock to the boil.

Heat 50ml of the olive oil with the butter in a large heavy-based saucepan, add the orzo and stir vigorously until the colour of the orzo deepens to golden. Add the rice to the pan and stir so the rice is well coated with oil. Add the cinnamon, sage, lemon peel and stock, and bring to boil. Put the lid on, turn the heat down and cook until the liquid has all gone, about 15 minutes. Keep warm.

Fry both types of the nuts separately in the remaining olive oil until golden brown. Drain off the fat and tip the nuts on a kitchen cloth to absorb the fat.

When cold, chop them roughly. Put in a small mixing bowl with the chopped parsley, add some seasoning and mix well. Set aside.

Prepare the chicken livers: trim any little sinews from the livers and cut them in half if quite large. If the liver has a little green bag, the gall bladder, attached to it, it should be removed with the sharp point of a knife, as the liquid inside is inedibly bitter.

In a mixing bowl, roll the livers in the flour mixed with some salt until well coated. Heat the oil and butter in a heavy-based frying pan large enough to hold all the livers in a single layer, or use two smaller frying pans. Empty the livers into a sieve and shake well to get rid off any excess flour.

When the fat is foaming-hot, season the livers and fry over a high heat for less than 2 minutes, turning once, until nicely browned on the outside but still pink and juicy in the centre. Lift them out of the pan and tuck them in the sieve.

Wipe out the frying pan with a wad of kitchen paper and add half the olive oil and the shallots. Sauté for 2–3 minutes over medium heat until lightly browned. Add the Mavrodaphni and reduce the liquid until it is just a glaze. Add the tomatoes and the remaining olive oil, and bring to the boil for a few minutes. Remove the frying pan from the heat and adjust the seasoning.

To serve, turn the rice into a large serving bowl, place the livers on the top, pour over the sauce and finally sprinkle with the mixture of nuts and parsley.

Anchovy fillets with Santorinian capers and dill

The Lesvian equivalent of sushi! This is one of the earthly pleasures that always lives up to the desire that drives me towards them. It is a long journey to Mytilini, the big 'city' and main harbour on the island of Lesvos. Most of the anchovies sold there during the season in July and August come from the bay of Kaloni, which is more like an inland sea. The locals say that the trick is to eat them before they eat their way through the cans!

What Lesvos is most known for is its ouzo. One of my favourites is Barbayiannis ouzo, from the village of Plomari, 'ouzo capital of the world', where the inhabitants – perhaps not unexpectedly – have a reputation for being just a little bit crazy.

Visanto is one of Santorini's traditional wines, made from sun-dried grapes, which after the harvest are spread out on flat roofs for 10 to 12 days.

Serves 6

30 salted anchovy fillets, rinsed
40g capers
20g caper leaves, if available
400g tomatoes, roughly chopped
50g rocket, finely chopped
75g dill, finely chopped

For the shallot vinaigrette

80ml extra virgin olive oil
40ml Visanto wine
Juice of 1 lemon
50g shallots, finely chopped
1 garlic clove, finely chopped
Salt and coarsely ground black
*   pepper*

To serve

warm bread
warm soft-boiled eggs

Coarsely chop the anchovy fillets and put them, together with all the other salad ingredients, into a medium-sized salad bowl.

In another bowl, mix all the ingredients for the shallot vinaigrette.

Pour the dressing over the salad, season and toss well.

Spread it over warm bread and eat it with warm soft-boiled eggs.

Leg of lamb with chickpeas, seasonal vegetables and mint

I usually make this dish with lamb shanks but, as so often happens on these voyages, the day I chose to cook it we weren't able to find shanks, but I did manage to get a hold of a 2-kilo leg of lamb, to which I gave exactly the same treatment – except, of course, I didn't boil it first and just used water instead of the stock.

Serves 6

500g chickpeas, preferably peeled
1 leg of lamb, about 2kg, or 6 lamb
*   shanks, each about 350g*
5 bay leaves
150g leeks, well rinsed and finely
*   diced*
150g carrots, peeled and finely
*   diced*
150g tomatoes, skinned, deseeded
*   and coarsely chopped*
100g potatoes, peeled and cut into
*   very small cubes*
100g courgettes, finely diced
100g rocket, coarsely chopped
100g watercress, trimmed and
*   coarsely chopped*
250ml olive oil
Juice of 1 lemon
100g mint, finely chopped
Salt and freshly ground black
*   pepper*

Soak the chickpeas overnight.

If using the lamb shanks, first put them in a large pan, cover them with water, bring to the boil, skim off any scum and add some salt. Simmer them with the lid on for 40 minutes.

Whether using the shanks or preparing to use the leg of lamb (see paragraph 7), in another large heavy casserole, gently heat 100ml of the oil and sauté the leeks for 2–3 minutes with the lid on. Stir in the drained chickpeas, carrots and bay leaves.

Only if using lamb shanks, transfer them to the casserole and add 2 litres of the lamb shank stock, bring to boil, turn the heat down, cover and simmer, stirring occasionally for 40–50 minutes. (Keep the remaining stock aside, just in case you need it later on.)

Preheat the oven to 180°/gas 4.

Whether using the lamb shanks or preparing to use the leg of lamb (see next paragraph), add the tomatoes, potatoes and courgettes to the casserole. The idea is to start first with the vegetables that take longer, remembering to season with salt as you add each vegetable.

If using the leg of lamb, put it on top of the bean and vegetable mixture at this point. Whether using the shanks or the leg of lamb, cover and cook in the oven for 1–1½ hours, checking from time to time and adding more lamb stock or water if necessary.

Remove from the oven, discard the bay leaf and stir in the rocket, watercress, chopped mint, the remaining olive oil and the lemon juice. Allow to rest for 15 minutes before serving.

4

Crustaceans Drowning in a Greek Coffee Pot of Aegean Water

We consider seafood to be the quintessence of the Greek 'eating soul' and love it for all its romantic affiliation

with the sea. Most of the stories you hear about the denizens of the Aegean from the local island fishermen

blend to form a colourful world full of partly fictitious, partly true escapades while they were at sea,

imbued with the marvels of the watery world. What always lingers in my mind as I listen to their tales

is that when they are not at sea the stories are always laden with a heavy nostalgia.

Prawns with apricot feta salad

Serves 6

*1kg raw prawns, unpeeled and
  heads on, rinsed and dried*
400g soutzouki sausage

For the marinade

6-8 saffron threads
50ml white wine
250ml olive oil
Juice and grated zest of 2 lemons
*1/2 teaspoon fennel seeds, toasted in
  a dry pan and ground*
5 garlic cloves, finely chopped
1 bunch of basil, finely chopped
*1 bunch of flat-leaf parsley, finely
  chopped*

For the salad

500g apricots, halved and stoned
100ml olive oil
1 red onion, thinly sliced
2 cucumbers, peeled and sliced
200g watercress, trimmed
200g feta
*1 tablespoon sesame seeds, toasted
  in a dry pan*

For the dressing

150ml extra virgin olive oil
*50ml aged red wine vinegar (if
  unavailable, use balsamic
  vinegar)*
1 teaspoon Dijon mustard
*Sea salt and freshly ground black
  pepper*

At least one hour ahead or preferably the day before, make the marinade: using a pestle and mortar, crush the saffron and pour the wine over it, stir well and keep warm. In a large bowl, whisk together the oil and lemon juice, then add the lemon zest, fennel, garlic, basil and parsley. Stir in the wine mix. Toss the prawns in the marinade and leave for at least an hour, preferably overnight.

Remove the prawns from the marinade about 30 minutes before cooking and preheat the grill.

Prepare the salad: brush the apricot halves with oil and sprinkle with salt and pepper. Place on/under a hot grill and lightly colour the apricots. When warm, pull from the grill and set aside. In a bowl, gently toss the onion, cucumber, watercress, feta and apricots.

To make the salad dressing, whisk the olive oil, vinegar and mustard together in a medium bowl until well blended.

Place the prawns on the grill pan with a piece of the soutzouki sausage under each, and place under the grill. Cook for 3–5 minutes or until the prawns turn pink and look firm but still moist, basting them frequently with the marinade.

Pour the dressing over the salad and mix gently. Sprinkle over the sesame seeds and serve while the prawns are still hot.

Lemony rice pilaf with cockles

Serves about 6

1.5kg cockles

Salt and freshly ground black
   pepper

200ml olive oil

100g fennel, finely diced

400ml white wine

1kg leeks, white and light green
   parts only, very well rinsed and
   cut into 1cm-thick slices

2 lemons, peeled, thinly sliced and
   any pips removed

500g short-grain rice

100g chilled butter, diced into
   small pieces

Rinse the cockles several times with plenty of running cold water. If any of the cockles are open – even slightly – don't use them. If you are preparing this dish while you are on any of the islands or on the boat, after the cockles are washed, put them in a pot full of clean sea water, otherwise in salted water.

In a heavy-based casserole, gently heat 100ml of the olive oil with the fennel and cook for couple of minutes or so, without allowing it to colour. Add the cockles and cover with the lid. Shake the casserole and, a minute later, add the wine. Turn the heat to high and continue cooking, still covered, letting the alcohol evaporate. As soon as the cockles are open, remove the casserole from the heat, drain the liquid through a fine sieve into a clean bowl and set it aside. Take all the cockles out of their shells, discarding the ones that haven't opened.

In a large casserole, place the leeks, 700ml water, some salt, the sliced lemons and half of the remaining olive oil. Bring to the boil, lower the heat and braise with the lid on until the leeks are almost crunchy. Add the cockle stock and the rice, stir well and cook until the rice is just tender and the liquid is absorbed, about 12-15 minutes. Do not stir during the cooking and have some hot water ready just in case the pilaff gets too dry.

Fold in the butter with a wooden spoon, adjust the seasoning, add in the shelled cockles and serve, drizzled with the remaining olive oil.

The sea urchin forms a large spiky ball resembling a hedgehog. It loves living in sheltered bays, and above all seeks out cooler water with high salinity, therefore the Aegean and waters around the Ionian Islands are among its preferred homes.

The sea urchin has two openings: the anus in the centre among the spines; and the mouth, known as 'Aristotle's lantern', found on the underside and having five teeth but no spines. This is where the edible parts of the animal are found: the five bright orange sex glands, often confusingly called the roe, and the liquid that surrounds them. The glands form five 'tongues', arranged in a star shape. I like sea urchins when they have 'put on their jewellery', that is when they have collected seaweed on their spines and look more like little Christmas trees.

Avgotaraho is poor man's caviar. In the Ionian Sea, along the west Greek sea line from Igoumenitsa to Pilos, the water in many spots gets very deep very soon offshore. There grows, mates and lives the female kefalos (grey mullet), the fish from which avgotaraho (bottarga) is produced. The roe is massaged by hand to eliminate air pockets, then dried and cured in sea salt for a few weeks. The result is a dry hard slab, which is coated in beeswax for keeping.

Sea urchins, avgotaraho, fennel and kohlrabi salad

Serves 4–5

48 sea urchins
150ml extra virgin olive oil
Juice of 2 lemons
300g fennel with its flowers
About 25 breakfast radishes,
  trimmed
500g kohlrabi, peeled
1 bunch of basil, picked and finely
  shredded at the last moment
1/2 bunch of mint, picked and finely
  shredded at the last moment
100g avgotaraho
Sea salt and coarsely ground black
  pepper
Warm bread to serve

If you are inexperienced, opening sea urchins can be time-consuming, so open a bottle of wine! If you are in Greece, buy from any fishing shop a pair of sea urchin pliers. Position the inside blade of the pliers around the sea urchin's mouth area, squeeze the handle and the sea urchin will separate into two parts. When you have opened all of them, use an espresso spoon to scoop out the five fleshy orange tongues. They simply smell and taste of the sea.

In a bowl, whisk half of the olive oil, half the lemon juice and the sea urchin juices until well amalgamated. Add the sea urchin eggs and keep the bowl aside.

Ideally using a mandolin or, if you do not have one, a very sharp knife, thinly slice the fennel and the radishes, and put them in a mixing bowl. Grate the kohlrabi, add it to the bowl with the herbs and toss the salad with the remaining olive oil and lemon juice.

Slice the avgotaraho on top, season with black pepper and finally pour the sauce with the sea urchins over the salad. Mix the salad well and serve straight away, with plenty of warm bread.

Crab meat on grilled sourdough bread with dill dressing

Picking the meat out of a crab is a labour of love, but there are fishmongers who sell freshly picked white or brown crab meat. Always look for non-pasteurized crabmeat.

A live crab should feel heavy for its size and smell fresh, with no hint of ammonia, and should look lively, otherwise do not risk it as, once dead, the flesh deteriorates in no time. At their best between April and December, a good crab should yield up to 50 per cent of its body weight in meat.

The water in which you cook the crab should be salted: either use the failsafe way which is when an egg floats on the surface of the water then there is enough salt; or use about 3 tablespoons of salt per litre of water. Cooking times vary according to size. Drop the crab in the boiling water and, when the water comes back to the boil, simmer for 10 minutes for the first 500g, and add a further 10 minutes for every extra 500g.

To remove the meat, first twist and break off the claws, then set the crab on its back on a chopping board and, as you hold it firmly down with one hand, give the underside a sharp thump with your other hand, pushing up with your thumbs, to remove the underbody and legs from the shell.

Pull away the grey gills, or 'dead man's fingers' from the crab's body and discard. Scoop out the creamy brown meat from inside the shell, which is the most intensely flavoured meat but looks unappetizing. Pull apart the crab's body and, using a skewer, remove the white meat from the two outer pieces of the body. Do the same with the legs. To crack the claws, hit them two or three times with a rolling pin in order to get at all the white meat.

Serves up to 6

Up to 1.3kg crab, male or female
1 tablespoon Dijon mustard
1 tablespoon thyme honey
150ml extra virgin olive oil
300g tomatoes
1 bunch of spring onions, trimmed
*   and thinly sliced*
200g purslane, trimmed
1 bunch of dill, trimmed and finely
*   chopped*
Salt and freshly ground black pepper
2 lemons
At least 6 slices of sourdough bread,
*   each about 2cm thick*

Bring plenty of salted water to boil and drop the crab(s) in. Once cooked (see previous page), with a slotted spoon, remove it and let it cool. Take off the upper shell and extract the meat as described. Keep brown and white meat refrigerated in separate bowls.

In a mixing bowl, dissolve first the Dijon mustard and honey in the juice of 1 of the lemons and then whisk in half of the olive oil until mixture is smooth and amalgamated. Keep it in the fridge.

Skin, deseed and finely dice the tomatoes. In a serving bowl, combine the white and brown crab meat with the tomatoes, spring onions, purslane, dill and three-quarters of the dressing. Season the mixture with salt and pepper, and mix well.

Preheat the grill. Brush the bread slices with the remaining dressing and grill until warm on both sides. Serve and leave everyone to place as much crab salad as they want on the top. Have a few more slices of bread ready for grilling, if needed, and cut the remaining lemon into wedges.

Plaki means a dish, normally of fish or beans, baked in the oven, with tomatoes, garlic, parsley and onions. Evia, or Euboia, is a very tricky island to navigate – even the intrepid corsairs found it so – as it has few lighthouses and is subject to very strong northerly winds and difficult currents. Having sailed these waters, you might reach one the oldest towns on Evia, Kymi, built high above the sea. Here during late August and September you will find the best figs.

Lobster plaki with a fig and manouri salad

Serves 4–6

2 live lobsters, each about 1kg
200g manouri cheese (if unavailable, use mascarpone)

For the tomato sauce

200ml extra virgin olive oil
150g shallots, finely chopped
5 garlic cloves, finely chopped
150g leeks, trimmed, well rinsed and finely chopped
300g tomatoes, skinned, deseeded and finely chopped
100g plump, moist ready-to-eat sun-dried tomatoes, finely chopped
50ml white wine
Juice of 2 oranges
Salt and freshly ground black pepper
1 bunch of flat-leaf parsley, picked and finely chopped
½ bunch of mint, picked and finely shredded
Juice of 1–2 lemons

For the grilled fig and manouri salad

6 figs
2 tablespoons aged red wine or balsamic vinegar
2 teaspoons thyme honey
Sprinkling of ground cinnamon
50ml extra virgin olive oil
Juice of 1 lemon

Cut the lobsters in half lengthwise with a sharp knife or cleaver, remove and discard the gritty stomach sac. If there is any coral, put it, together with the green stuff (the tomalley or liver – a great delicacy), into a strainer over a bowl, sprinkle with salt and reserve.

Make the tomato sauce: in a heavy-based pan, heat 3 tablespoons of olive oil and sauté the shallots, garlic and leek gently until translucent. Add the chopped fresh and sun-dried tomatoes, the wine and orange juice. Season, cover and simmer for 15 minutes.

While the sauce simmers, start preparing the salad: snip the tiny stem end off each fig and cut in half lengthwise. Mix the vinegar, honey and cinnamon together in a mixing bowl. Add the figs and gently toss to coat. Let sit while you preheat a medium grill and the oven to 220°C/gas 7.

When ready, grill the figs (reserving the marinade in the bowl), for a minute or so per side, or until grill marks appear. Do not over-cook or the figs will become too mushy. Transfer the figs to a plate.

To the reserved marinade, add the remaining olive oil, lemon juice, salt and pepper, whisking well to completely incorporate. Add the manouri to the plate of figs and pour over the marinade.

Finish the tomato sauce by adding the parsley and mint, adjust the seasoning, stir and remove from the heat.

Put the lobster halves in a roasting tray. Spread sauce liberally over the flesh and roast for 10 minutes or just until the flesh turns white.

While the lobsters are roasting, push the coral and liver of the lobster through the strainer and beat the purée with the remaining olive oil and the lemon juice. When the lobsters are cooked, drizzle the liver and coral vinaigrette over their flesh.

Serve immediately with the salad.

Mussel soup

Serves up to 6

3kg mussels
150ml olive oil
2 bunches of spring onions, trimmed and finely chopped
6 garlic cloves, thinly sliced
1.5kg ripe tomatoes, skinned, deseeded (reserving the juice, about 500ml after straining) and chopped
1 bunch of basil, leaves and stalks separated and both finely chopped
Sea salt and freshly ground black pepper
Rusks or sourdough bread, to serve

Scrub the mussels well and remove any beards, then rinse very well. If any mussels are open, or even just about, and don't close if tapped, don't use them.

Heat the olive oil in a large heavy saucepan, add the spring onions and garlic, and cook gently until lightly coloured. Add the chopped tomatoes and diced basil stalks, and keep cooking over higher heat, stirring, until the tomatoes break up and reduce to a thicker sauce, about 15 minutes.

Add the mussels (if the saucepan is not big enough, do this in 2 batches) and the reserved tomato juice. Cover tightly and cook over a high heat until the mussels have opened. Remove from the heat, leave to cool for 15 minutes, then remove all the mussels from their shells, discarding the ones that haven't open.

Add the mussels to the runny tomato sauce and mix gently. Adjust the seasoning and then add the chopped basil leaves.

I like the way the locals eat it, at room temperature, poured over rusks in a bowl and drizzled with some more extra virgin olive oil. If you can't find rusks, then slice some sourdough bread, brush the slices with some garlicky butter, put under the grill for couple of minutes and pour the soup on top of them.

Razor clams flambéed with ouzo

On its own, pastrouma – a cured beef fillet rubbed with fenugreek and a myriad of spices – can be quite robust

Serves 6

300g new potatoes, peeled
Salt and freshly ground black
*   pepper*
200ml olive oil
1 bunch of spring onions, trimmed
*   and thinly sliced*
¹/₂ bunch of fresh lemon thyme,
*   picked*
3 bay leaves
4 garlic cloves, peeled and crushed
24 razor clams
100ml ouzo
50g mustard leaves, finely shredded
100g rocket, trimmed and finely
*   chopped*
100g plump, moist sun-dried
*   tomatoes, roughly chopped*
50g flat-leaf parsley, finely chopped
100g pastrouma (if unavailable,
*   use pastrami), rind removed*
*   and thinly sliced*
Juice of 2 lemons

Boil the potatoes in salted water. When cooked, drain, then cut the potatoes in half and set aside in a bowl.

In a large saucepan over medium heat, combine 100ml of the olive oil, the spring onions, thyme, bay leaves and garlic. Cover and cook for 2-3 minutes, then add the razor clams with the ouzo and cover tightly again. Steam the razor clams until the shells have completely opened and the alcohol has evaporated.

With a slotted spoon, remove the clams from the saucepan and reserve. Add to the broth in the pan the halved cooked potatoes and the mustard leaves with seasoning to taste. Mix well and simmer for a few minutes, until the sauce becomes slightly thicker.

Remove the saucepan from the heat, add the rocket, tomatoes, razor clams, parsley and the remaining olive oil, and toss well.

Arrange the pastourma slices on top, drizzle over some lemon juice and serve.

Cooking while sailing is like a game of chess, and the day that I will have learned all the rules, I shall then manage to finalize a day's menu before sunrise! Everyone was still asleep while Jason and I were experiencing the magnificence of a great majolica net extended from the stern of a nearby fishing boat. The fishermen spread out on it the samples of the early-morning hard work: squid, *marida* (tiny silver Aegean whitebait, coming in hundreds per kilo, even as adults) and shellfish. With the smiles of mermaids on their faces, they arranged the catch on the deck of the fishing boat in a certain order and with great respect. We wake up the others with two buckets full of still-live squid and *marida*.

Fried squid with a salad of artichoke, potato and anchovy

Serves 5

12 young artichokes

4 lemons

10g whole black peppercorns

Salt and freshly ground black pepper

500g small new potatoes, peeled

100ml aged red wine vinegar

50g anchovies canned in oil,
drained and coarsely chopped

100g rocket, coarsely chopped

150g dry sourdough bread or rusks,
roughly crushed

For the Visanto wine and shallot dressing

200ml extra virgin olive oil

50ml Visanto wine (a sweet pale
red Greek wine; if you can't
find any, use a good sweet
sherry)

50g shallots, finely chopped

For the squid

1.5kg squid

750ml olive oil or any other good-
quality vegetable oil for frying

about 400g plain flour for dredging

Sea salt and black pepper

Lemon wedges

Prickly artichokes are such a disarming puzzle of the chopping board. Cooking and eating artichokes is like crossing the threshold to another somewhat familiar world, from which joy and satisfaction are never banished. Pick artichokes that are dark green and have tightly closed leaves. Avoid ones that are dry-looking or turning brown. If the leaves appear too open, then the choke is past its best.

First prepare the artichokes for the salad: peel off their tough outer leaves and trim off the top third of the artichoke. Peel the stems.

Put the juice of 4 of the lemons in a medium-sized casserole, add the artichokes, the juiced lemon halves, whole peppercorns and a tablespoon of salt. Cover the artichokes with cold water. Bring to the boil and simmer for 10 minutes, or until the artichokes are tender but still firm when the tip of a knife is inserted in them. With a perforated spoon, remove the cooked artichokes from the simmering water and place them, top sides down, on towels in the fridge and leave until completely cold and dry.

Drop the potatoes into the artichoke water and bring to the boil. Simmer until cooked but not falling apart, about 15 minutes. Drain in a colander and set aside to cool and drip dry completely.

Make the Visanto wine and shallot dressing: in a small bowl, whisk together the oil, wine, chopped shallots and seasoning to taste. In a small pan, bring the aged red wine vinegar to the boil; reduce the heat and let it simmer for 3–5 minutes, or until the vinegar has reduced to a syrupy consistency that will coat the back of a spoon.

Scrape out any fuzzy centres from the artichokes and peel away the remaining inner leaves (these could have some nice fleshy tips which someone might enjoy as a snack or starter) to reveal the hearts.

Prepare the squid: holding each body firmly, twist the head and pull it away from the body without breaking the ink sac. Cut the tentacles from the head just below the eyes. At the centre of the tentacles there is a small beak, squeeze to remove it, discard and wash the main body inside and out to get rid of any sand and tissue. At the same time, wash the tentacles too. Drain well. If you have managed to save the ink, freeze it for other recipes.

Pour the oil into a large heavy-based frying pan and get it quite hot, but be careful that it does not get hot enough to smoke. Place some flour and a bit of salt into a fine sieve, add a few squid pieces, shake to cover and then shake off any excess flour. One at a time, gently drop them in the hot oil. Cook in batches, leaving enough room in the pan to turn over the pieces and fry until the squid turns golden and crispy, about 3–4 minutes. Remove from pan and drain on paper towel. Repeat until all squid is done. Season well.

While the squid are frying, transfer the artichokes and the potatoes to a large salad bowl. Add the chopped anchovies, Visanto wine and shallot dressing, aged red wine reduction and chopped rocket. Toss very gently. Sprinkle the crushed rusks or dry sourdough bread over the top.

Serve the salad straight away, together with the hot squid and lemon wedges.

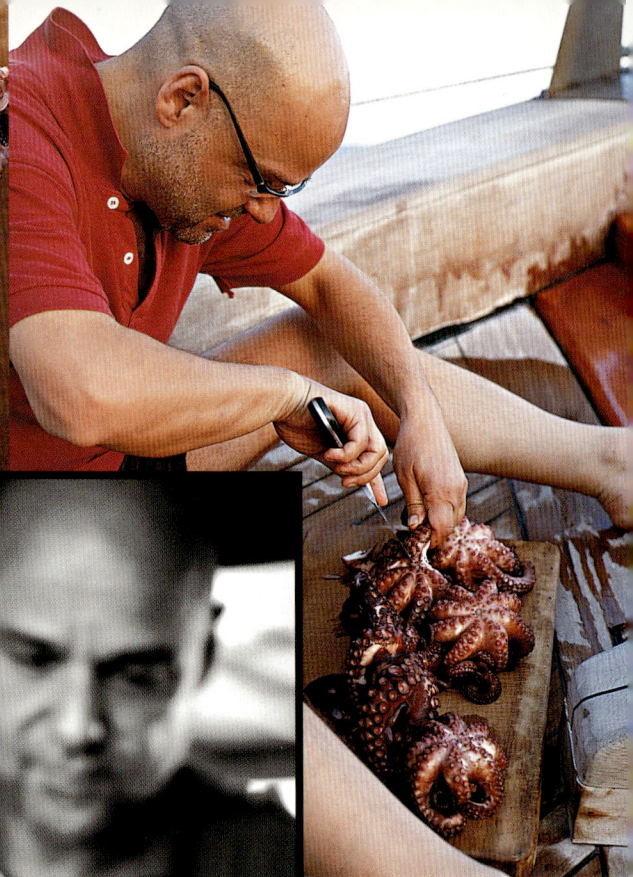

Octopus with rosemary and garlic

Antiparos is the island of the surprises, as you can walk to the two neighbouring islands, Diplo and Kavouras. Here is the best sea bottom in the whole Aegean for octopus. The Mediterranean octopus, distinguished by having two rows of suckers on its tentacles while its North Atlantic relatives have only one, better suits this particular treatment.

Serves 6

2kg octopus, washed well and left whole with the head attached
75g garlic, peeled but left whole
125ml extra virgin olive oil
½ bunch of rosemary

Preheat the oven to 120°C/gas ½.

Place a large heavy bottomed casserole over heat and let it get hot. Put the whole octopus in the casserole and put the lid on. You don't need to add any liquid as it will produce its own.

Put the garlic cloves in a small saucepan, add the olive oil and rosemary, cover with the lid and bake in the preheated oven for at least 40 minutes to 1 hour, until tender.

When ready, separate the garlic from the oil and set both aside, discarding the rosemary. Remove the octopus from the casserole, place it in a large mixing bowl, cover it with cling film and set aside.

Meanwhile, boil the octopus liquid to reduce it to about 300ml. The reduced liquid will be quite concentrated.

Put the whole garlic cloves in a liquidizer and turn the motor on for couple of minutes to purée it. Start pouring in the olive oil in a thin steady stream. When you have used all the olive oil, add the reduced octopus juices in the same way. At the end, you should have very well amalgamated, thick garlicky sauce.

Portion the octopus into manageable pieces and place these in a large mixing bowl. Pour the sauce on the top and mix well. Preferably, let the flavours get friendly for an hour at room temperature before serving.

Stuffed Cuttlefish

The island of Karpathos combines attractiveness, remoteness, wilderness and what we Greeks call 'dialogicos euphoria' (dialogical euphoria). When the tourist season finishes – just as the cuttlefish season begins – part of the island, especially the village of Olympos, seems caught in a time warp, as the locals speak a dialect which has traces of the ancient Dorian Greek and women in traditional embroidered dress rule the village. They bake their bread in communal outdoor ovens; the village's windmills are still medieval and the feel is like an open, living Aegean garden, where architecture, ethnology, linguistics and musicology interact.

When the local women and men get together, they express feelings of sympathy and love for each other without violating the social correctness of their culture, through dialogical singing and dancing. You put across your euphoria in public and negotiate boundaries of some real or imaginary personal concerns. Being with the locals for a few days, I managed to fine-tune some of my thoughts concerning loneliness, the self-reflexive condition of being remotely part of our world, separation, expression and embryonic desire.

Serves 6

4kg cuttlefish
400ml olive oil
500g shallots, thinly sliced
Salt
300g medium-grain rice
100ml white wine
200g apricots, stoned and coarsely
*   chopped*
150g pine nuts
100g flat-leaf parley, finely chopped
2 bunches of spring onions, trimmed
*   and finely chopped*
1 cinnamon stick, ground
3 cloves, ground
1 nutmeg, grated
400g tomatoes, skinned, deseeded
*   and coarsely chopped*

Cleaning cuttlefish can be messy as these marine creatures have an ink sac that ruptures easily. Don't be put off by the colour as the ink rinses off easily! Rinse the cuttlefish, remove the head with the tentacles and discard the intestines. Pull out the cuttlebone, remove the skin and rinse well. Remove the eyes and beak from the head, but leave the head attached to the tentacles and pull or rub off the skin from the tentacles, or as much as will come off easily.

Very gently heat 100ml of the oil in a large heavy-based pan and add half of the shallots. Cover the pan and sweat slowly for about 15 minutes, until the shallots are softened but not coloured.

Add the cuttlefish heads and tentacles, and keep cooking gently with the pan uncovered until the cuttlefish juices have evaporated. Rest the cuttlefish on a chopping board and let it cool down.

Bring to boil 500ml of water and add some salt.

To the same pan in which the cuttlefish were cooked, add the remaining shallots with 100ml of the remaining olive oil, cover and sauté gently over medium heat. When the shallots look soft and pale, add the rice, turn the heat up to medium and, with a wooden spatula, stir until the rice is well coated in the oil and onions. Pour in the wine and keep stirring until it has totally evaporated.

Add the apricots, pine nuts and the 500ml hot salted water. Cover the pan and simmer for 10-13 minutes, or until the rice is tender. Remove from the heat and add the chopped parsley and spring onions.

Chop the cooked cuttlefish tentacles and head, add to the pan with the rice, mix well and finally adjust the seasoning.

Score the cuttlefish body diagonally each way so to make a diamond pattern. Spread some of the rice stuffing over the non-scored flesh, roll and close up by securing the two ends with cocktail sticks.

Pour 100ml of the remaining olive oil into a straight-sided casserole and sprinkle in the cinnamon, cloves and nutmeg. When the aromatics begin to sizzle and smell, arrange the stuffed cuttlefish in the casserole, open-side down. Brown for about 3-4 minutes, then carefully turn and brown all over. Add the tomatoes and the remaining oil, shake the casserole well, cover and braise for 10-12 minutes, until the cuttlefish is tender.

If the sauce is still slightly runny when the cuttlefish is ready, then remove the cuttlefish with a perforated spoon, turn the heat up and cook the sauce for a few more minutes uncovered. Turn the heat off, put back the cuttlefish and let it rest for 15 minutes, then serve.

Pan-seared pinna

The is THE Aegean delicacy – with a drop of olive oil and some lemon juice you have just arrived on a Greek island. Pinna is a large bivalved shellfish that can grow up to 75cm. Shaped like a Parma ham, it is also known as the fan shell mussel, due to the thin fragile triangular shells that are a light yellow-brown to a darker brown in colour. A series of raised fine sculptured lines run across the shell, while the interior of the shell is glossy and, when the shell dries out, it becomes very brittle and cracks.

Serves 4

4–12 pinna, depending on their size
Well-seasoned white flour, for
  dusting
knob of butter
Drop of olive oil
2 lemons, halved

If you can find them, open them in the same way as you would scallops: remove and clean the white part of the pinna, slice it in half, and just before you are about to cook them, roll them in well-seasoned white flour, shake them well to loose any excess flour and drop them into a hot heavy-based frying pan with a knob of butter and drop of olive oil and sear them for just a minute.

Transfer them to a platter and eat them while hot, with some lemon juice drizzled over them. If you see them while you are swimming, they also taste amazing raw, as long as you have a knife and a lemon with you!

CHAPTER

5

Cooking While Heading into a Thunderstorm

The title has a metaphorical meaning; you prepare yourself in advance for the rainy days. In our Aegean archipelagos, there is a myriad of tiny islands which are cut off from the mainland and each other when the weather conditions worsen. The sense of feeling abandoned is evident, but the locals keep the stocks of food soaring, and that is one of the things that make these islands irresistibly charming.

Back on the boat, the remote possibility I had been clinging to that I would have been able to cook through the storm had evaporated. In such circumstances I enjoy the kind of comfort food that this situation favours, which is very tasty but with no frills. What makes it very exciting is having had the foresight to stock up the cupboard and fridge with very accessible satisfying treats: chickpea soup, mackerel fillets smoked in mountain tea from the islands, shrivelled olives, pickled aubergines, raw spring onions, tarama paste, herring roe, indigenous Cycladic garlic, cheese, olives and Seville orange preserve – and enough Metaxa brandy to loosen our tongues.

Revithosoupa (chickpea soup)

Chickpeas appear in our daily diet in a hundred guises! You smell them everywhere in wonderful casseroles and soups in combination with spices, vegetables or meat.

Serves 6

500g chickpeas (preferably the peeled ones)
200g onions, diced
200g leeks, trimmed, well rinsed and finely chopped
250ml olive oil
1 teaspoon cumin seeds, lightly toasted in a dry pan and then ground
5 garlic cloves, crushed
6 bay leaves
6 sage leaves
100g pine nuts
Juice of 2 lemons
Salt and freshly ground black pepper
300g beetroot leaves, trimmed and finely shredded (if not available, use spinach)

Soak the chickpeas in cold water overnight.

In a heavy-based casserole, sauté the onions and leeks in 100ml of the olive oil, with the lid on, over low heat, until glistening. Add the cumin, garlic and herbs, and give them a quick stir.

Drain the chickpeas, put them in the casserole with double the quantity of water as that of the chickpeas and cover with the lid again. Turn the heat down and simmer until soft, about 1½ hours, skimming from time to time during the first 30 minutes.

About 30 minutes before the chickpeas will be cooked, colour the pine nuts gently in a tablespoon of the remaining olive oil. Add them to the soup, together with the lemon juice. Adjust the seasoning and remove the casserole from the heat when the chickpeas are tender. Add the beetroot leaves and mix in, together with the remaining olive oil.

This dish actually improves with standing, so it can be cooked the day before, but always add the raw beetroot leaves just as you are about to serve it. If there are any leftovers, drain off any excess liquid and next day turn the beans into a chickpea bourekia (see overleaf).

Bourekia me revithia

If you have any leftovers from the Revithosoupa on the preceding pages, you can use them to make little pies called bourekia. Such filo pastry rolls are generally made with a filling of cheese or seasonal vegetables. These are a variation on ones I tried during Lent, on the island of Amorgos.

Serves 4–6

About 250g leftovers from the Revithosoupa on pages 116–17
2 tablespoons extra virgin olive oil
2 medium-sized tomatoes, skinned, deseeded and well chopped
about 100g any good tasty cheese, grated
Salt and freshly ground black pepper
About 12 sheets of filo pastry
A little melted butter
1 egg beaten with a little milk

To the leftovers add the olive oil, tomatoes, cheese, with salt and black pepper to taste. Set aside.

Preheat the oven to 180°C/gas 4.

Cut the filo pastry sheets into rectangles about 10x20cm. Brush one sheet with a little melted butter and egg wash, put another sheet on top and brush in the same way. Then put a tablespoonful of filling at one end of the sheet. Fold over the sides and nearest end of the filo and roll the pastry lengthways into a sausage shape. Make 5 more rolls in the same way.

Bake the rolls in the preheated oven for about 15–20 minutes until nicely coloured.

Lentils with a touch of spices from Mytilini

This is one of the simplest and best meals you can prepare on the boat, and it is full of restorative powers.

Serves 6

250ml olive oil
1/2 teaspoon cumin seeds, toasted in a dry pan and ground
1/2 nutmeg, grated
1/2 teaspoon ground cinnamon
1 teaspoon sweet paprika
3 whole cloves, ground
200g banana shallots, finely chopped
150g leeks, trimmed, well rinsed and finely chopped
500g lentils
4 bay leaves
4 sage leaves
70g flat-leaf parsley, finely chopped
40ml red wine vinegar
Salt and freshly ground black pepper

To serve

plenty of warm bread
olives
cured anchovies

Bring 2 litres of water to the boil.

In a large heavy-based casserole, mix 100ml of the olive oil and the spices. Place over a low heat and let the aroma of the spices reach your nose. Add the shallots and leeks, cover and sweat for no more than 10 minutes, until soft and pale.

Add the lentils, bay leaves and sage tied together. Stir for a few minutes to mix them well, then pour the boiling water into the casserole. Bring to the boil, turn down the heat and simmer for about 1 hour. At this point do not add any salt as it will harden and discolour the lentils.

Towards the end of the cooking time, remove the herbs and bay leaves, add the remaining olive oil and finish cooking.

Remove from the heat and let the soup rest for 10 minutes. Add the chopped parsley and red wine vinegar, with seasoning to taste. Serve with plenty of warm bread, olives and cured anchovies.

Run out of flour for bread?

Kefalotiri, a traditional hard cheese, is the first cheese produced at the start of the new season, ensuring the milk that is used to make it is taken after the goats and sheep are weaned. Similar to Italian pecorino, it is firm and dry in texture and has a slightly sharp finish.

Trahana is the humblest of pastas, an old recipe for preserving dairy produce during lean months. Best results are obtained when it is used in a ratio of 1:5 to 1:6 of liquid.

Serves about 6

500ml full-fat milk
4 sage leaves
100g yoghurt
100g trahana (see page 15)
100g feta
100g kefalotiri cheese
2 tablespoons fresh mint leaves, chopped
2 tablespoons fresh flat-leaf parsley, chopped
2 tablespoons olive oil, plus more for frying if necessary
50g butter
2 eggs
Freshly ground black pepper

In a heavy-based saucepan, warm the milk with the sage leaves over medium heat. Pour in the yoghurt, whisk and add the trahana. Reduce the heat, add the feta and simmer, stirring occasionally, until the trahana is tender and creamy.

Add the kefalotiri, herbs, oil and butter, and stir well while you are still over the stove, until the trahana looks like a runny porridge. Remove from heat and let it cool for 15 minutes.

Add the eggs and mix well. If the trahana has thickened a lot, loosen it up with some more hot milk, and turn it out on a rectangular baking sheet lined with baking parchment, like an enormous flat cake. Leave to cool completely.

When cold and set, cut it into slices as you need it and either bake, or shallow-fry or grill. Season with pepper and serve with any leftovers, dips and salads.

Elephant bean casserole with orange and fennel

Greek elephant (or giant) beans stand out for their huge size (there are on average about 40 per 100g), their kidney-like shape and their pure white colour. They originate from the area of the lakes of Prespes in Florina, Northern Greece. The combination of the microclimate and the soil composition of the specific cultivation area are responsible for their wonderful flavour. In 1994, Greece acknowledged the elephant bean of Florina as of Protected Designation of Origin (PDO) status and the European Union accorded them Protected Geographic Indication (PGI) status. Use butter beans if you can't find them.

Serves 6

500g elephant beans, soaked in cold water overnight, preferably for 24 hours, but not in the fridge
4 bay leaves
Juice and peel of 4 oranges
200ml olive oil
100g onions, coarsely chopped
100g carrots, coarsely chopped
100g parsnips, coarsely chopped
100g celery, coarsely chopped (use the lighter green stems and leaves)
3 garlic cloves, peeled and left whole
300g tomatoes, skinned, deseeded and coarsely chopped
200g fennel, including the flowers if any, trimmed and finely diced
1 head of curly endive, trimmed
Salt and freshly ground black pepper
1/2 bunch of flat-leaf parsley, finely chopped
1 bunch of basil, shredded at the last moment

Drain the beans, put them in a heavy-based casserole and cover with fresh water, about double the volume of beans. Add the bay leaves and orange peel, cover and bring to the boil. Turn the heat down and simmer, still covered, until just becoming tender, 45–60 minutes. As they simmer, uncover from time to time and skim.

Preheat the oven to 160°C/gas 3. In a heavy frying pan, heat 100ml of the oil and sauté the onions, carrots, parsnips, celery and garlic until soft. Add the tomatoes and cook for about 5 minutes. With a hand blender, purée the vegetables.

Discard the orange peel from the beans, then add the vegetable mixture to the casserole, together with the orange juice, fennel and curly endive. Stir gently and lightly season with salt.

Place the casserole in the oven and bake for 2–3 hours or until all flavours have been absorbed by the beans and they are tender. Check the dish occasionally and, if more liquid is needed, add a little hot water from a kettle. Do not add cold water, as it will make the bean skins tough. At this stage the beans should look fairly wet.

Add the remaining oil, the parsley, basil and fennel flowers if you have them, adjust the seasoning and stir well. Allow to cool with the lid left on. The beans can be eaten at room temperature too.

October and the first sausages

The Greek Islands offer a unique range of sausages (loukaniko).
This one is a blend of the first autumnal pork mixed with the local
goat meat, orange peel and lots of crushed red pepper flakes.

Makes about 12–14

*500g boneless shoulder of pork,
    untrimmed and coarsely minced*
*500g boneless shoulder of goat,
    untrimmed and coarsely minced*
*500g leeks, trimmed, rinsed well
    and finely chopped*
3 garlic cloves, finely chopped
Grated zest of 2 oranges
1 teaspoon fennel seeds
1 teaspoon crushed red pepper flakes
8 sage leaves, finely shredded
*100ml retsina (if not available, use
    any red table wine)*
*Salt and freshly ground black
    pepper*
*About 3 metres pork or lamb
    sausage casings, well rinsed*
Olive oil for cooking
Knob of butter

Well ahead, ideally a couple of days before you need them: in a
large bowl, toss together thoroughly all ingredients apart from the
oil and butter. Cover with plastic wrap and leave refrigerated for at
least 4 hours, preferably overnight, for the flavours to blend.

Open one end of the sausage casing, fit it over the tap in your kitchen
sink, and place the remainder of the casing in a bowl under the tap.
Turn the tap on gently to wash out the casings, which usually are
sold clean and salted. Now you are ready to start stuffing. Insert
one end of the casing over the funnel attachment of a standard mixer
or meat grinder. Push the casing on to the length of the funnel (it will
squeeze and fit) and leave about 5cm dangling from the end.

Turn the mixer on and, as the mixture begins to flow into the
casing, it will push the casing off the funnel. As the meat reaches
the dangling open end, tie this end with a double knot. Have a
large bowl to catch the sausages and make several very long
sausages. Be careful that the sausage stuffing enters the casing
continuously and evenly and that no air bubbles develop. Do not
overstuff the casing, as it will burst there and then or during
cooking. If you get air bubbles, it is better to cut the sausage at that
point and start a new one by tying the end off.

When all the mix is in the casing, divide it into portions about
10cm long and refrigerate for up to 3 days. Fresh sausages develop
more flavour if allowed to rest in the fridge for at least 24 hours.

When you want to cook them, remove them from the fridge and
keep at room temperature for 30 minutes. Glaze a heavy frying pan
with olive oil and butter and place over medium heat. When the
pan is hot, add the sausages and sauté, turning and browning
slowly, until cooked but still moist, about 15 minutes. For grilling,
prepare a barbecue or preheat an overhead grill on low for 10
minutes. Grill the sausages for about 15 minutes, turning frequently.

Froutalia from the island of Andros

Each island is responsible for stocking the produce and staples it makes and grows. The daily conversations that are likely to be heard are about the pantry, clay jugs, wooden barrels and the hanging produce. Household economy is the principal law that rules the majority of the Greek Islanders.

Froutalia is an everyday food on Andros that changes according to the season and the seasonal produce. There are therefore many different recipes but for all of them the frying pan, good olive oil and local eggs as common elements.

Serves 6

200g potatoes, peeled and cut into about 2cm dice, rinsed and left in cold water
Salt and freshly ground black pepper
100ml olive oil
6 sausages
1 bunch of spring onions, trimmed and coarsely diced
2 roasted peppers (from a good-quality jar will be fine), cut into thin strips
50g flat-leaf parsley, trimmed and finely chopped
¹/₂ teaspoon ground cinnamon
10 eggs, lightly beaten
Lemon wedges, to serve

Cook the potatoes in gently boiling salted water until tender, about 10 minutes. Drain and allow to cool.

Heat half of the olive oil in a large heavy-based frying pan and add the sausages. Cook gently over medium heat, turning and browning for 10 minutes.

Add the spring onions and potatoes, and sauté for 5 more minutes. Add the peppers, parsley and cinnamon, and stir gently.

Turn the heat down to very low, remove the sausages from the pan, cut them into small pieces and return them to the pan.

Add the eggs, tilt the pan slightly so that the entire bottom of the pan is covered and season with black pepper. When the omelette begins to set, run a spatula along the rim of the pan and slightly underneath, allowing the uncooked egg to run. Cook for another minute or so.

Place an inverted plate atop the pan and invert pan and plate together quickly (not as tricky as it sounds). Then slip the omelette back into the pan and cook the other side for another minute.

Transfer the froutalia from the pan to a flat warmed serving platter and serve hot with lemon wedges.

The first of the season's apples for Freddie and the other kids

Serves 6 easily

50g local sultanas
50g dried apricots, stoned and finely chopped
50g dried figs, finely chopped
50g dried prunes, stoned and finely chopped
Juice of 1 lemon
3 tablespoons thyme honey
1 teaspoon ground cinnamon
½ bunch of mint, picked and shredded at the last moment
6–8 apples
About 80g butter

Preheat the oven to 200°C/gas 6. In a medium-size bowl, mix the dried fruits with the lemon juice, honey, cinnamon and mint. Set aside.

Wash the apples but do not peel them. With a corer, clear out the centres of the apple (you may need to remove a little more than just the core to give enough space for the filling) and place them on a heavy-based baking sheet. Squeeze the fruit mixture into the cavities together with some butter. Don't worry if it doesn't all fit in, as the butter will melt and cover the top of the apple, giving it a shine as it bakes.

Bake for about 30–40 minutes; poke gently with the tip of a sharp knife to check they are ready, remove and serve hot or at room temperature.

Spicy tomato and walnut dressing for cold meat or fish

As I was preparing my shopping list for the food and location photography, I was introduced to 'tomato perasti' which is almost like a tomato passata, and the one I use here was produced by Alexandros Gousiaris, who lives in the small village of Ilias in Central Greece.

The tomatoes, a local variety known for their aroma, grow on his family farm where they are selected by hand for ripeness. They are then washed, pressed and strained to produce a delicious base for soups, sauces and casseroles. No other ingredients are added, just simply good tomatoes, stored in your cupboard, as they will be the cooking 'blood' for your winter cooking.

You can make it yourself and all you need is 500g of great ripe tomatoes. Remove the skins and seeds, and whisk the flesh in a blender. Put it in cheesecloth and hang it until you get a thick tomato pulp, which will weigh about 100g. If you have outdoor space, use the seeds for the next year's production!

Serves 4-6

White of 1 egg
1 teaspoon soft brown sugar
½ teaspoon cumin seeds
1 level tablespoon sea salt
100g walnuts
400g tomato perasti
2 mild red chillies, deseeded and finely chopped
1 garlic clove, finely chopped
Juice of 1 orange
200ml extra virgin olive oil

Preheat the oven to 160°C/gas 3.

In a mixing bowl, whisk the egg white with the sugar, cumin and sea salt, and fold in the walnuts.

Cover a heavy-based baking sheet with baking parchment, spread over the 'showered' walnuts and bake for about 20 minutes until quite dry but not too coloured. Make sure you keep an eye on them until they are done, so that they don't burn.

In the goblet of an electric blender combine all the remaining ingredients with the roasted walnuts and blend to a smooth thick purée, stopping the motor and scraping the sides as needed.

Keep refrigerated and use liberally over any leftover cold meat or white fish.

Sweet tomato perasti with aged red wine vinegar and cinnamon

Tomato perasti is the Greek equivalent of passata – see opposite. Intensely dark-coloured vinegar with a delicate fruity flavour and sweet, clean and lasting aftertaste is produced from Corinthian grapes grown in the Kalamata region of the Peloponnese. Aged in wooden barrels, this is a Greek alternative to balsamic vinegar.

Serves 4–6

330g jar of tomato perasti (see opposite; if unavailable, use good-quality passata)

2 tablespoons aged red wine vinegar

120ml extra virgin olive oil

1 tablespoon thyme honey

1/2 teaspoon ground cinnamon

Put all ingredients in a food processor and blend until the sauce is well amalgamated. Keep it refrigerated and use it as a dip, or over eggs, or as salad dressing.

Gourounaki kokkinisto (pork casserole)

On the boat, and with the direct exposure to a strengthening Aegean summer winds, known as the meltemi, it makes it easier to enjoy this dish by warming up some flat bread, adding a few flakes of the meat and some yoghurt cheese (see page 133), wrapping it and eating it warm, while you appreciate the rugged beauty of the Aegean wilderness.

Serves 6–8

150ml olive oil

2kg shoulder of pork, on the bone

250g onions, finely chopped

150g carrots, peeled and finely chopped

6 garlic cloves, peeled and left whole

8 sage leaves

1/2 bunch of thyme

500g tomato perasti (see pages 128–29; if unavailable, use a good-quality passata)

1 bottle of red wine

Salt and red pepper flakes

2 tablespoons aged red wine vinegar

Heat half the olive oil in a large heavy wide-bottomed casserole. Add the pork and sear for about 5–8 minutes over medium heat, until slightly crusty all around.

Add the chopped vegetables, garlic and the two herbs tied together. Sweat for 5 minutes without letting them colour. Stir in the tomato and simmer for a few more minutes, then pour in the wine, add some salt and red pepper flakes, and bring to the boil. Turn the heat down to a simmer, cover and cook for about 1$^{1}/_{2}$ hours. When the meat is ready, it will come off the bone and break readily into big flakes.

Stir into the sauce the remaining olive oil and vinegar, adjust the seasoning if needed and take off the heat.

Be sure to choose a shoulder of pork that is not completely a stranger to fat. The fat in the shoulder will melt slowly and help protect the pork intimately as it cooks, keeping the meat moist and flavoursome.

Yoghurt-based cheese with rosemary

You can serve this drizzled with olive oil and sprinkled with black pepper for a snack or add it into salads.

Serves 6

*2 litres full-fat milk, preferably
  sheep's or goats'*
30g rosemary, picked
3 tablespoons plain yoghurt
1 level tablespoon salt

Bring the milk with the rosemary in it to just under boiling point and then pour the milk into a large deep mixing bowl. Let the milk with the rosemary cool to about 42°C.

Remove any skin that has formed on the surface, pass the milk through a sieve and discard the rosemary. Whisk in the yoghurt, cover with a cloth, place in a warm, draft-free environment and leave for 24 hours.

Mix the salt into the runny yoghurt and empty it into clean cheesecloth. Tie the ends together and suspend it in the fridge over a deep bowl for about 3 days, depending on your desired firmness.

CHAPTER

6

Meats that Melt in the Mouth

The terrain of the Greek Islands is friendlier to goats and sheep than cattle, and thus beef dishes tend to be almost a

rarity by comparison. The freedom of the animals to climb, explore and feed from the scarce vegetation on the local

dry rocks makes the meat irresistibly tasty, and local long slow cooking techniques ensure that they also instantly

dissolve on the palate, sometimes even saving you the trouble of having to chew it!

Lamb mastella

On the island of Sifnos we had lamb mastella. Instead of roasting the lamb, it is cooked in a clay pot. Until that day, I had always been a great fan of lamb on the spit, but I changed my mind when I tried it this way. Lamb mastella is awesome. It is not what you cook, it is how you cook it.

Serves about 6

2kg leg or shoulder of lamb, left
    on the bone
1 bunch of thyme
1/2 bunch of marjoram
2 whole garlic heads, unpeeled
300g carrots, peeled and left whole
300g button onions, peeled and
    left whole
100ml olive oil
Salt and coarsely ground black
    pepper
2 lemons, peeled, thinly sliced and
    pips removed
200ml red wine

Preheat the oven to 230°C/gas 8.

Put the lamb in a clay pot or heavy-based casserole big enough to allow you to get some herbs, vegetables and garlic around the meat. Strip the thyme and marjoram off their stems and chop finely, setting the stems aside. Stir the olive oil into the chopped herbs to make a spreadable slush, and then crumble in some salt and crushed black pepper. Massage this seasoned oil all over the meat. (I always feel that there is something quite pleasurable about doing this.) Then cut the whole heads of garlic in half and tuck them under the lamb with the lemons, carrots, onions and stems from the thyme and marjoram. Pour in the wine and seal well with a lid.

Cook in the oven for 20 minutes, then turn the oven setting down to 160°C/gas 3 and continue roasting until the meat is done as you would like it. Lamb needs about 40 minutes per 1kg, plus the initial twenty minutes, so for a 2kg piece you should start checking after the meat has been in the oven for 1 hour 20 minutes. This will give you a medium lamb, still juicy and pink in the middle.

Remove the meat from the oven, discard the herb twigs, pull out the garlic cloves from their skins, and leave the lamb to rest for 20 minutes on a serving platter in a warm draft-free place. Strain the cooking liquid into a bowl, spoon off any fat, roughly chop the vegetables and lemon, add the roast garlic squeezed out of its skins and mix with the liquid in the bowl. Adjust the seasoning if needed.

To serve, carve the meat in thin slices (it will be more likely that the meat will be falling off the bones anyway) and serve with the chopped vegetables.

Chicken soup trahana

Is it the boat or the compass or the trahana soup that makes me turn? Regardless of how it is cooked, you need an 'instant-read' meat thermometer, and the chicken is cooked when the internal temperature of a thick part of the bird reaches the very low 70s°C. Do not rely on the colour of the juices to determine whether is done. Most chickens we buy nowadays are slaughtered while still under-aged, so their bones are short of sufficient calcium. Thus blood can escape from the bones, creating an illusion that it is undercooked. Internal temperature is the only reliable way to judge.

Serves about 6

1.5kg chicken

200g button onions, peeled and left whole

1 medium-size head of fennel, trimmed and cut in half

2 lemons, quartered, plus more to serve

1 bunch of flat-leaf parsley, left whole

Salt and freshly ground black pepper

100g carrots, peeled, trimmed and left whole

1 cinnamon stick

6 bay leaves

150g trahana (see overleaf)

100ml extra virgin olive oil

Fill the cavity of the chicken with the onions, fennel, lemon, parsley, salt and pepper, then truss it. This allows the entire bird to cook evenly.

Select a saucepan into which the chicken will fit comfortably. Place the chicken in the pan, add the carrots, cinnamon and bay leaves, and season with the salt and pepper. Fill the saucepan with water until it covers the chicken, about 2.5–3 litres, and then bring to the boil. Turn the heat down to a simmer, cover and cook the chicken for 1 hour but, after the first 45 minutes, start checking the temperature. Obviously, if you haven't got a meat thermometer, use the traditional way of checking, that is when the chicken's juices run clear. Once the chicken is cooked, transfer to a large bowl with a cup of the cooking liquid, cover and let it cool down.

With a slotted spoon, fish out the carrots, bring the stock to the boil and add the trahana. Simmer, adding more water if needed, until the trahana is cooked. Remove the onions, fennel, lemons and parsley from the chicken's cavity, remove the meat from the bones and shred it into chunks. Stir everything apart from the lemons gently back into the pot with the trahana.

Serve immediately after adjusting the seasoning with freshly ground black pepper and a good drizzle of fine olive oil. I have a few extra lemons on the side just in case you prefer the soup a bit more sour, as I do. The finished dish should be juicy chunks of chicken swimming in a shallow matrix of aromatic soup, not flavourless rags of chicken flesh in a solidified trahana!

Make your own trahana

Is it not amazing that yoghurt, milk and cracked wheat can produce something so basic but yet delicious?

Serves about 6

450ml full-fat milk, warmed
300g yogurt
1 tablespoon lemon juice
600 cracked bulgur wheat
30g plain flour
40g semolina
12 eggs, lightly beaten
1 tablespoon salt

The day before: in a large bowl, combine the milk, yogurt and lemon juice. Cover and let stand for 24 hours to sour.

In a large casserole, mix the bulgur wheat, flour and semolina, and then blend in the soured milk mixture and the eggs. Bring to the boil very gently, reduce the heat and simmer for 10 minutes, stirring with a wooden spoon. Season with salt and let it sit for 30 minutes.

Preheat the oven to 120°C/gas ½. Place large spoonfuls of the mixture on a baking sheet, place in the preheated oven and let dry for 6–8 hours.

When dry, place pieces in a food processor and pulse for a second to break into pebbles. Stored in a container in a cool dry place, it will keep for several months.

Into the Aegean and Eastern Mediterranean pot have gone the eating habits of the Middle Eastern, Turkish and Greek cultures. Some of them have been modified through the years but trahana has remained exactly as it was in the beginning – called *trakta* by the ancients.

Grilled kid with green tomatoes

Serves about 6

1.5kg kid, off the bone (if
unavailable, use lamb instead),
cut into walnut-sized pieces.
150ml olive oil, plus more for the
baking sheet
generous sprinkling of Greek
mountain oregano
1 teaspoon ground cinnamon
3 garlic cloves, finely chopped
150ml Mavrodaphni or port
2 lemons
1 tablespoon Dijon mustard

For the green tomatoes

1kg green tomatoes, cut into 1cm
slices
2 good-sized lemons
2 tablespoons brown sugar
Breadcrumbs from tasty bread
1 bunch of dill, finely chopped
Salt and coarsely ground black
pepper
2 bunches of spring onions, trimmed
and finely chopped
100g butter, cut into tiny pieces

Well ahead, in a bowl large enough to hold the meat, mix together the olive oil, oregano, cinnamon, garlic and Mavrodaphni. Add the meat and stir to coat evenly. Grate the zest of all lemons, juice them and stir both zest and juice into the mixing bowl. Cover and refrigerate for 2–3 hours, preferably overnight.

Light the barbecue and preheat the oven to 200°C/gas 6.

To prepare the tomatoes, grate the zest of the lemons into a small bowl and mix together with the sugar, breadcrumbs, dill, salt and pepper.

Oil a baking sheet and arrange a layer of tomatoes on it, sprinkle with the spring onions and then with the breadcrumb mix. Continue layering in this way until all tomatoes are used up. Spray the top layer of tomatoes with lemon juice and finish with a final layer of breadcrumb mix. Dot with the butter and bake for 30 minutes.

As you are about to cook the meat, thread the pieces on wooden skewers that have been well soaked in water. Season with salt and pepper.

Fold the mustard into the marinade and whisk until you get a well-amalgamated sauce.

Let the fire burn down to almost a white ash before you begin. I always put my hand just above the grill to check the heat; if I can hold it for over 5 seconds, then the fire is ready to be used, otherwise the flame is too hot and will burn this delicate meat. Set the skewers on the grill and cook to your desired doneness, turning the skewers and basting them with the marinade frequently for even cooking. Greeks (me excluded) generally cook their meat until it is well done, but in any case, both of us serve them hot and with the warm salad of tomatoes!

Flomaria (hilopites) with milk-roasted baby goat

Serves about 6

150ml olive oil
2kg baby goat shoulder on the bone
(if unavailable, use lamb)
Salt and freshly ground black pepper
300ml full-fat milk
6 sage leaves
4 bay leaves
1 cinnamon stick
3 cloves
2 tablespoons yoghurt
750g tomatoes, skinned, deseeded
and coarsely chopped
About 500ml chicken stock or water
500–750ml boiling water
500g hilopites (see opposite)
Plenty of freshly grated kefalotiri
cheese (if unavailable, use
pecorino)

Preheat the oven to 160°C/gas 3.

In a medium-sized enamelled, cast-iron casserole or any other heavy-based casserole, heat the oil, reserving 1 tablespoon. Season the goat with salt and add it to the casserole. Cook over moderately high heat, turning once, until lightly browned on all sides, about 10 minutes or more. Spoon off the excess fat in the casserole. Add the milk, sage, bay leaves, cinnamon stick and cloves to the casserole and bring to the boil. Turn the heat off.

In a separate mixing bowl, whisk a couple of ladlefuls of the milk with the yoghurt and pour the mix back into the casserole. Cover with baking parchment and a lid, and transfer to the oven. Braise for about 1 hour 20 minutes, turning the meat occasionally and spooning the cooking liquid over it from time to time. The meat is done when it's tender enough to pull off the bone.

Add the remaining tablespoon of oil to the casserole along with the tomatoes. Stir in the stock or water, bring to boil, season with salt and pepper and simmer over a low heat until the sauce has more body and is reduced to about 800ml, about 1 hour.

Remove the meat from the casserole and set aside in a warm place. Sieve the cooking juices and measure the volume; you should have about 150ml. Stir them into the reduced tomato sauce and adjust the seasoning if necessary.

Preheat the oven to 250°C/gas 9.

Pour the sauce into a baking dish, add 500–750ml of boiling water, just enough to give a soupy consistency, stir in the hilopites and bake for 15–20 minutes, stirring occasionally with a fork.

Pull the meat off the bones and add it in with the hilopites, mix in well, adjust the seasoning, grate over some kefalotiri and serve hot, passing more cheese separately.

Flomaria is a kind of homemade egg noodle. A staple in the Limnian household, it is usually made from basic raw materials produced or stored by the household, such as wheat, pulse flour, milk, eggs, water and salt, rolled into either long flat ribbons or little square pieces of varying thicknesses. Flomaria can be found at Greek grocers under the name '*hilopites*' or Italian '*quadrettini*'.

Rabbit and peas hotpot from the island of Halki

I have several blue-tinted memories etched into my mind and Halki is one of them. As the boat swings around the horseshoe-shaped harbour, you meet Nimborio, the only settlement on the island which does not have the typical Greek Aegean island architecture but instead sports a selection of two- and three-storey neoclassical houses. If you like nothing better than to sit on a quiet beach with a good book and a cold drink, giving free rein to your thoughts, then Halki is the place for you! Halki for me is a place of restoration, preservation, relaxation and imagination.

Just remember if you try to cook this recipe while in Halki that the local rabbits need more cooking, not like the fattened plump bunnies you come across back home.

Serves about 6

1 rabbit, about 1.2–1.5kg, jointed
    into 8 pieces
200ml red wine vinegar
1.5kg fresh peas in the pods
Pinch of sugar
About 300g plain flour
Salt and freshly ground black pepper
200ml olive oil
1kg button onions, peeled but
    leaving root ends intact, so they
    won't disintegrate while cooking
300g tomatoes, skinned, deseeded
    and grated
2 cloves
2 bay leaves
1 head of garlic, sliced across in
    half to halve each clove
1 bunch of lemon thyme
300ml chicken stock
1 bunch of dill, finely chopped
100g kefalotiri, grated (if
    unavailable, use pecorino)

The night before: put the rabbit pieces to marinate in a bowl with the vinegar and some water to cover.

The next day, shell the peas, put them in a saucepan and add enough water just to cover them. Add the sugar, bring to the boil and then simmer for about 15 minutes. Drain and set aside.

Preheat the oven to 150°C/gas 2. Drain off the marinade and pat the rabbit pieces dry. Season the flour and use to coat the rabbit portions. Heat 3 tablespoons of oil in a large, deep frying pan over medium heat. Add the rabbit and brown until lightly golden on both sides. If the pan isn't big enough to hold them all comfortably, brown in batches (you may need an extra splash of oil).

In a straight-sided casserole, mix the reserved peas, the onions, tomatoes, cloves, bay leaves and garlic in 3 tablespoons of the olive oil. Add the rabbit and lemon thyme, pour over just enough stock to cover the rabbit and, if it doesn't, add a little water. Cover and cook in the oven for 1½ hours, but do check the tenderness of the rabbit after the first 60 minutes.

When done, drizzle over the remaining oil, sprinkle over the dill and some kefalotiri, and serve hot, with plenty warm crusty bread.

Christmas in Syros – Mosxaraki kapama (a kind of veal stew)

I am not that sure whether this great festive veal stew originates from Syros, where it is often served at Christmastime, but come and smell it. Don't interrupt your reading with too many questions about the recipe, as it's likely that I won't know the exact answer and even if I try to answer them I will mess it up! The only 'secret' I can share with you is 'don't fuss, just let it cook'.

Serves about 6

2kg veal shoulder
4 mandarins, peeled and
then juiced
About 500ml chicken stock
60ml olive oil
50g butter
5 carrots, peeled and left whole
6 red onions, peeled and quartered
3 leeks, trimmed, well rinsed and
roughly chopped
Peel of 2 oranges, preferably Seville
2 cinnamon sticks
3 cloves
1 bunch of parsley, finely chopped
Salt and freshly ground black pepper
300g tomato perasti (see page 128)
or good-quality tomato passata
300ml red wine
Lots of watercress, to serve, if you
like it

Arrange the meat in a large bowl and rub all over with the mandarin juice. Allow it to stand while you carry on with the preparation.

Preheat the oven to 160°C/gas 3 and bring the stock to a simmer.

Place a roasting pan over a medium heat and gently warm the oil and butter. Sweat the vegetables in this until they just soften. Cuddle up the veal with the vegetables, citrus peel, cinnamon, cloves and parsley. Season, add the tomato perasti, wine and simmering chicken stock. Cover with a lid or kitchen foil and bake for 1½–2 hours.

It will perhaps look fairly simple for a festive dish, but do serve it warm with plenty of roughly chopped watercress and any other Christmassy delicacies.

Lamb baked in paper with seasonal vegetables

Unable to find any string on board the boat, we extemporized with crumpled kitchen foil to secure the paper. Although the removal of the paper to let the lovely aromas escape can be quite theatrical, this is a very simple dish, but worthy of the best china and glints of light from a few candles. After the initial wows, we weren't hanging around for any more words.

Serves about 6

1 leg or shoulder of lamb
4 garlic cloves, thinly sliced
Grated zest and juice of 2 lemons
250g new potatoes, peeled and
*   cut in half*
1 bunch of spring onions, trimmed
*   and coarsely chopped*
300g fresh peas, shelled
250g tomatoes, skinned, deseeded
*   and coarsely chopped*
4 tablespoons olive oil
1 bunch of mint, leaves picked
1 tablespoon chopped fresh aromatic
*   oregano*
Salt and coarsely ground black pepper
100ml red wine
150g kefalotiri cheese (if
*   unavailable, use pecorino)*

Preheat the oven to 180°C/gas 4. All over the meat make deep incisions with the tip of a small knife and insert slices of garlic and some seasoning into them. Rub the meat with lemon juice and zest.

Combine the potatoes, spring onions, peas and tomatoes in a large mixing bowl. Pour the oil into the bowl, sprinkle in the mint and oregano, add some seasoning and mix well.

Tear off a large sheet of parchment paper (do not use foil as it will prevent the meat from browning) and double it up so to avoid any leaks. In the centre of the sheet, pile all the ingredients and then nest the lamb in the middle. Drizzle over the wine and any leftover juices from the mixing bowl, squeeze the cheese in between the lamb and vegetables, gather up the sides of the paper, twist the top and tie with heatproof string (or crumpled foil).

Place the parcel in a lightly oiled heavy-based roasting tray, add a tiny puddle of water and cook in the oven for 1½–2 hours.

Allow to rest out of the oven for 20 minutes or so, open and serve on a large warmed platter.

Roast pork loin on the bone with chopped seasonal salad and yoghurt

Syros, with its capital Ermoupoli, is often called 'Queen of the Cyclades', as it stands on a naturally amphitheatrical site, with neoclassical marble buildings and white houses cascading down to the port. On the hill above is the picturesque medieval settlement of Ano Syros, where we met Filomila, who not only sold us her pork, but came on board to cook it too! Yoghurt is traditionally made from fresh ewes' milk. My favourite producer is Roussas from Almyros, close to Volos on the mainland. His product is aromatic and sharp, and has a compact texture, with a fat content of about 7–10%.

Serves about 6

2kg pork loin on the bone
5 garlic cloves, peeled and crushed
    to a fine paste
1 tablespoon late-picked thyme
Sea salt and freshly ground black
    pepper
2 tablespoon olive oil
200ml Mavrodaphni (if
    unavailable, use a good port)
3 tablespoons yoghurt
50g pine nuts
1 bunch of parsley, finely chopped
150g purslane, coarsely chopped
Juice of 1–2 lemons
About 120ml good olive oil

For the seasonal salad

500g tomatoes, coarsely chopped
100g soft moist ready-to-eat sun-
    dried tomatoes
1 cucumber, peeled and diced
2 bunches of spring onions, chopped
2 peppers, deseeded and sliced

Pork loin is a cut from the centre rib area of the loin, made out of the loin eye and rib bones. With a sharp knife or blade, score lines through the skin and fat (not going as deep as to touch the eye), 1–2cm apart. Mix the garlic, thyme and seasoning in a bowl and rub it over the flesh but not the skin. Lay skin side up on a board and pat the skin dry. Sprinkle generously with sea salt and let stand.

Preheat the oven to 220°C/gas 7 and lightly oil a heavy-based shallow roasting tray.

Shake off excess salt from the pork, place on the roasting tray and roast for about 30 minutes or until the skin has blistered and crisped (promising signs for good crackling). Turn the oven down to 160°C/gas 3 and cook for another 80 minutes or so. After an hour, if you have a meat thermometer, start checking the middle of the loin temperature; when it reads 70°C, it is cooked, otherwise use your judgement. Take out and let rest for 15–20 minutes.

While the meat is resting, make the salad by mixing the ingredients.

Pour off excess fat from the roasting tray, place it over medium heat, add the Mavrodaphni, simmer and scrape any caramelized bits off the bottom. Add the yoghurt, stir and toss the salad, pine nuts and herbs gently into the juices, empty into a platter and carve the meat. Drizzle olive oil and lemon juice over everything.

CHAPTER

7

Dinner Time under the Night Firmament

Rather than the instinctive closing of windows and pulling down of blinds to seal out the night, being at sea

encourages you simply to stretch out and fix your eyes on the great arcing expanse over your head and to

feel the freedom of being a tiny speck in the great scheme of things. After such a theatrical prologue,

we all feel ripe for dinner!

Heli spetsiotiko (Pot-roasted eel, Spetses style)

The island of Sptetses was once called Pityousa, meaning 'full of pines' and then the Venetians decided to rename the island 'Isola di Spezzie' or 'island of fragrances', probably because of its many aromatic plants. When it comes to the delights of the stomach, I find the eel the king of the feast.

Serves 4–5

3 tablespoons olive oil

500g button onions, peeled and left whole

4 garlic cloves, peeled and left whole

100g plump moist ready-to-eat sun-dried tomatoes, roughly chopped

300g tomatoes, skinned, deseeded and roughly chopped

1 tablespoon thyme honey

1 bunch of thyme

3 bay leaves

1kg eel, skin on but fins trimmed away, cut into 5cm-thick pieces

1 bunch of flat-leaf parsley, finely chopped

1/2 bunch of mint, finely chopped

Salt and freshly ground black pepper

Grated zest and juice of 1 lemon

250g feta cheese, crumbled

Crusty bread, to serve

In a heavy-based casserole big enough to accommodate all the ingredients, heat 2 tablespoons of the oil and gently cook the onions and garlic to a golden brown colour.

Add both kinds of tomatoes, the honey, thyme and bay leaves, and simmer for 15–20 minutes, until the sauce begins to thicken. Preheat the oven to 180°C/gas 4.

Add the pieces of eel to the sauce and stir in the parsley, mint and salt and pepper to taste. Sprinkle with the lemon juice and roast in the preheated oven for about 20 minutes. Strew with the crumbled feta and lemon zest, and shake the dish, so the cheese submerges into the sauce a little. Cook for about 10 minutes more.

Serve with plenty of crusty bread on the table.

Bonito and Santorinian spring vine leaf rolls with olive oil and lemon juice

Serves 4–5

About 40 vine leaves

500g thick bonito fillets

For the marinade

2 garlic cloves, peeled

1 teaspoon fennel seeds

Sea salt and coarsely ground black pepper

250ml olive oil

About 50ml lemon juice

50ml red wine

100g plump moist ready-to-eat sun-dried tomatoes, pounded to a paste

1 bunch of spring onions, very finely chopped

1 bunch of flat-leaf parsley, finely chopped

Soak the vine leaves for about 30 minutes in two changes of boiling water to remove the strong tang of the brine and make them softer. Drain and pat dry.

Make the marinade: using a pestle and mortar, crush first the fennel seeds and then the garlic with some sea salt until smooth. Transfer this fennel/garlic paste to a mixing bowl and whisk in all but a tablespoon of the olive oil, together with the lemon juice and wine. Add the tomatoes, spring onions and parsley, mix well and leave the marinade in the fridge until it is needed.

Cut the bonito into pieces that will fit in the frying pan, heat the pan over medium heat, drizzle the reserved tablespoon of olive oil and then put in the bonito with its skin facing downwards. To avoid the bonito curling upwards, lightly press on it with a spatula and make sure that the skin is cooked. Then lightly cook the other side so that it just changes its colour and the centre of the fish is still red and juicy.

Cut the fish fillets into pieces about 4x2cm. Depending on the size of the vine leaves, flatten out one or two vine leaves (if you are using 2 they should overlap) and place the bonito piece at one end. Roll the leaf around the fish firmly until the long side is entirely wrapped.

Place it in the bowl with the marinade and finish rolling all the pieces. Let the fish cool down in the marinade for up to 30 minutes.

Serve with the Mandarin, olive and sweet red onion salad with sesame seeds on page 161.

As we left the island of Pano Koufonisi, which lies between the islands of Naxos and Amorgos and has only about 360 permanent inhabitants, we reached Kato Koufonisi. This sits alone, virtually uninhabited apart from fishermen and yachties, for the whole year, except for one day, the 15th August, when people from the neighbouring islands arrive to celebrate the Virgin Mary. As the sun was setting, and the boats started sailing away, I got the distinct impression that the island feels at its best left alone and touched only by the Aegean waves.

That day was also memorable for us catching several decent-sized bonito off the stern of the boat. And that evening we made these vine leaf rolls. Bonito gets mushy if not eaten within a day or so of being caught. Here the bonito flesh is quickly seared over direct heat and served at room temperature while the inside is still raw.

Fisherman's relish – rabbit spread

Observing the normal day of a fisherman, it is surprising how much of it is occupied by the act of eating. Food is taken seriously and is very strongly linked to every form of social entertainment. Perhaps paradoxically, this rabbit spread is one of the things they often snack on, as rabbit are so abundant on the islands.

Serves 6–8

2 tablespoon olive oil
½ teaspoon freshly ground fennel
*   seeds*
3 cloves
20 whole black peppercorns, ground
2 pinches of freshly grated nutmeg
1.5kg jointed rabbit
1 red onion, quartered
2 leeks, trimmed, well rinsed and
*   chopped*
4 garlic cloves, chopped
250ml milk
1 tablespoon brown sugar
250g butter
½ medium-sized cinnamon stick,
*   ground*
Pinch of red pepper flakes
Salt

Preheat the oven to 160°C/gas 3.

In a medium-sized ovenproof casserole, gently heat the olive oil with the fennel, cloves, pepper and a pinch of nutmeg, until the first aroma is released.

Then lightly colour the rabbit pieces in the casserole, remove and set aside.

Add the onion, leeks and garlic, and sweat them without letting them get too much colour.

Return the rabbit to the casserole, pour over the milk, add the sugar, cover closely and cook in the preheated oven until the meat readily leaves the bones, about 1 hour. Allow to cool.

Pick the meat from the bones and chop it very finely. Put the butter in a bowl and beat until creamy, adding the cinnamon, remaining tiny pinch of nutmeg, red pepper flakes, salt to taste and then the ground meat. Mix until homogeneous and very creamy. Turn the mix into a deep bowl and refrigerate.

Serve with warm bread, spicy radishes, purslane or watercress, or thinly sliced raw fennel dressed with olive oil and lemon juice.

Celeriac roots and watercress

This dish is a marriage of intense flavours that are part of the very fabric of life on the Greek Islands

Serves 5-6

1kg celeriac root, peeled, reserving all the trimmings, and cut into walnut-sized pieces, then kept under cold water

Salt and freshly ground black pepper

50g butter, cut into small pieces

100ml olive oil

200g watercress

1 bunch of spring onions, trimmed and finely chopped

Put about 300g of the celeriac trimmings in a medium-sized casserole with 500ml water and a very little salt. Cover with a lid and bring to boil. Turn the heat down and simmer until the trimmings are very soft.

Preheat the oven to 200°C/gas 6.

In a much larger casserole, melt the butter over a low heat and arrange the celeriac pieces in a snug circle in it. Colour them lightly and evenly all around.

Purée the drained cooked celeriac trimmings and spread them evenly over the browned celeriac. Add the oil and some salt, place the casserole in the oven and cook until the celeriac is tender but still retains a bit of crunch.

Remove the casserole from the oven. Shred all the watercress very finely, sprinkle it over the celeriac together with the chopped spring onions, put the lid back on and let cool down for about 10 minutes before serving.

The deep blue inlet of Vathi bay in Kalymnos ends at the charming village of Rina, which has one main street and a short collection of houses scattered into the valley, with an unexpected patch of lush vegetation. Vathi is the only place in the entire island with its own spring water supply, which is used for their citrus fruits, figs and very sweet grapes.

Mandarin, olive and sweet red onion salad with sesame seeds

Serves 4–5

250g mandarins, peeled and
   separated
1 Seville orange, peeled, pith
   removed and thinly sliced
30 Kalamata olives
30 natural cracked green olives
1 red onion, sliced (preferably with
   a mandolin)
1/2 bunch of mint, finely shredded
1/2 teaspoon fennel seeds, toasted in
   a dry pan and ground
1 1/2 tablespoons sesame seeds,
   toasted in a dry pan for about
   5 minutes
Touch of cinnamon
Pinch of red chilli flakes
75ml fine, peppery extra virgin
   olive oil
Some sea salt

Put the fruit, olives, onion, mint and seeds together in a large mixing bowl.

Mix the remaining ingredients well in a small bowl. Pour over the salad and gently toss together.

Pressed cockerel

Tsiknopempti is an annual ritual during carnival. It is a day of joy and of meat eating, and is always celebrated on a Thursday, eleven days before 'Clean Monday', the beginning of Orthodox Lent. In some areas this dish is traditionally served on Tsiknopempti.

Serves about 12

1 capon or a good-quality, free-range chicken, about 2.5kg
500g banana shallots, peeled but left whole
500g carrots
1 head of garlic, left whole
5–6 celery stalks, trimmed
Small bundle of herbs, such as flat-leaf parsley, thyme, bay leaves
Peel of 1 Seville orange
2 cinnamon sticks
Pressed juice from 250g apples
1 litre chicken stock
250ml red wine
Salt and freshly ground black pepper
100ml extra virgin olive oil
1 bunch of flat-leaf parsley, very finely chopped
Grated zest of 2 lemons

In a large casserole, put all the ingredients except the olive oil, bunch of parsley and lemon zest. Add the peel and juice of 1 of the lemons, and some cold water if there is not enough liquid to cover the bird. Bring to the boil and, as soon as it boils, turn the heat to low, cover with a lid and simmer for 50–60 minutes.

Take it off the heat and leave it to cool down. You should have a very moist bird that won't be tough. Drain the liquid into a smaller casserole, bring that to the boil and reduce it by about 60 per cent.

While the liquid is being reduced, discard all the citrus peel, herb bundle and cinnamon stick, and pick the meat off the bones. Slice all the vegetables, and sprinkle over half of the parsley, then add both to a large mixing bowl together with the chicken. Drizzle over the olive oil and stir together. Adjust the seasoning but keep in mind that the reduced stock should have a strong flavour.

Pour the half of the reduced liquid into the bowl and mix well. Layer the meat in a bread mould or, better, a terrine dish if you have one, lined with baking parchment. Pour the remaining liquid over the top, spread the remaining parsley evenly over, followed by the grated lemon zest, and allow to cool.

Cover with cling film, cut a piece of thick cardboard to fit on the top, add couple of jars to weight it and put it in the fridge for a day.

Turn the terrine out of its container and slice it with a very sharp knife. Savour this joyful texture with pickled onions and a Parsley and onion sauce (see overleaf).

Mussels, clams and anything else from the sea bottom, on the barbecue

The ratio of mussels to any type of clams depends entirely on you and your pocket. Make sure all crustaceans are alive and any soft-shell clams respond to your touch.

Serves 6

4kg mixed crustaceans
4 tablespoons olive oil
2 peppers, deseeded and finely chopped
1 chilli pepper, deseeded and finely chopped
4 garlic cloves, finely chopped
12 ripe plum tomatoes, skinned, deseeded and chopped, with their juice
1 tablespoon aromatic dry thyme
100ml tsipouro or grappa
200g feta, broken into pieces
Salt and coarsely ground black pepper

Prepare your barbecue to that perfect glowing moment and position the greased grill 15cm above the white coals. Preheat the oven to 200°C/gas 6. Rinse any dirt off the clam and mussel shells.

Heat the olive oil in a large ovenproof non-stick frying pan and sauté the peppers and chillies over medium heat. Add the garlic, tomatoes, oregano and tsipouro, cover and simmer for about 20 minutes.

Place the clams and mussels on the grill, making sure that they're not overlapping and are resting on the curved half of their shells so their juices won't escape. Start removing shells away from the heat when they open wide, about 3 minutes.

Scoop the meat out with a paring knife and put in a bowl with their juices.

Add the feta to the cooked peppers and tomatoes, and place the pan in the preheated oven, until the feta begins to melt. Just before removing from the oven, swirl in the mussels and clams with their juices. Season carefully (the feta is quite salty).

Eat while hot.

Parsley and onion dip from Syros

Serves about 12

250g stale bread, crusts removed
About 150g flat-leaf parsley, finely
*   chopped*
1 red onion, finely chopped
1–2 egg yolks
300ml extra virgin olive oil
Juice of 1–2 lemons
Salt and freshly ground black
*   pepper*

Pass the bread under cold running water to dampen it and then squeeze out the excess water.

Pulverize the parsley and chopped onion in the food processor, then add the bread, pulsing on and off to combine. Add the egg yolks and keep pulsing on and off a few times.

Slowly drizzle in the olive oil and lemon juice, alternating between each, pulsing until the mixture is smooth and creamy.

Season to taste with salt and pepper. Serve with warm crusty bread or as an accompaniment to grilled fish, meat and terrines. It will keep happily in the fridge for up to 3 days.

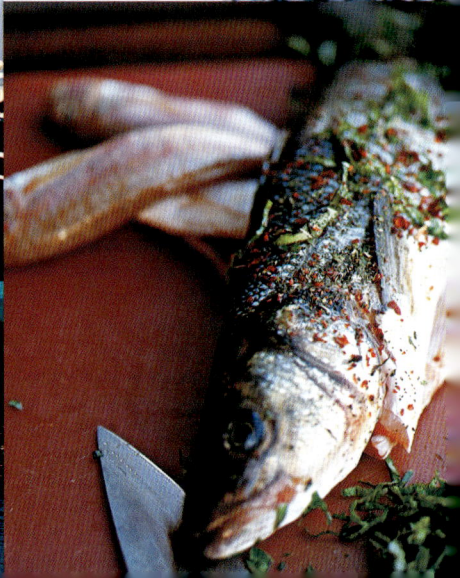

A few simple thoughts, which are not a 'real' recipe – grilled whole fish with fennel

The majority of the Islands don't have a fish market as such, as everyone just goes to the port each morning to wait for the fishing boats to return. Such almost-live fish have an utterly delicious flavour and texture that are almost impossible to find unless you live by the sea. The best fish is obviously the freshest fish, meaning no more than half a day old.

Whole fish
Salt and freshly ground black pepper
Fennel
Extra virgin olive oil
Lemon juice
Dried aromatic oregano or thyme, crushed

Preheat a griddle, grill or barbecue. Scrape off the scales from the fish but, if your fish is heavier than 2kg, leave the scales on to protect it from burning, as they need longer cooking time, but remove the skin before eating. Carefully slice into the stomach cavity to remove the intestines. Leave the head on if you can. Rinse well and pat dry. Lightly salt the interior and exterior of the whole fish.

Slice the fennel thinly lengthwise with a mandolin and, using only the tender inside part of the fennel, stuff it into the belly cavity with a few lemon slices. Set in a colander over a bowl and refrigerate until the griddle, etc is ready.

Brush the fish with olive oil and lemon juice, and sprinkle with oregano or thyme. Brush the griddle or barbecue with olive oil to prevent sticking, and cook the fish – using a hand-held rack makes life easier – turning often to grill evenly on both sides. It will be ready when the flesh flakes easily but still remains juicy inside.

Pull the skin off the top of the fish with the fingers, loosening it with a fork. Serve with olive oil and more lemon juice.

Slow-cooked veal cheeks

We're in Samothraki, one of the last truly 'virgin islands' of Greece, where modern civilization hasn't yet managed to ruin its natural beauty. Here the locals say if the meat has a face, eat it – yes, eat the face! Cheek meat is some of the most succulent meat there is, and when properly cooked melts into an 'archipelago' of meaty morsels, packed with a punch but still delicate in flavour. The mixed spices have a woody bouquet, and give a flavour that has a stimulating pepper bite but no single flavour dominates. Any you don't use will keep well in a clean dry place out of the light and can be used for all sorts of meat and poultry dishes.

Serves 4 plus

1.5kg veal cheeks
80ml olive oil
500g button onions, peeled and left whole
400g carrots, peeled and left whole
750g new potatoes, peeled
5 garlic cloves, peeled and quartered
300ml red wine
400ml chicken stock
Salt BUT NO BLACK PEPPER
1 bunch of flat-leaf parsley, finely chopped

For the mixed spices

1/2 teaspoon paprika
1 teaspoon red pepper flakes
1 teaspoon ground cumin seeds
1 teaspoon ground coriander seeds
4 cloves, ground
1/2 teaspoon ground fennel seeds
1 nutmeg, freshly grated
1 cinnamon bark, ground

Preheat the oven to 120°C/gas 1/2.

Trim the cheeks with a sharp knife to remove any silver skin; you'll end up with them about 40 per cent lighter in weight.

Mix the spices in a shallow bowl and drench the cheeks in 1 1/2 tablespoons of the mix. Roll and tie them well with butcher's string.

In heavy-based casserole, heat the oil gently and seal the meat rolls evenly, then remove them from the casserole to a plate.

In the same casserole, add all the vegetables and the garlic, cover and sweat them for a few minutes.

Transfer the cheeks back to the casserole and nestle them in between the vegetables. Pour in the wine, stock and some salt. Bring to the boil, cover and transfer immediately to the oven.

Cook slowly for 2 hours but start looking for the tenderness in the meat after 1 1/2 hours. When ready, the meat should be tender enough to pull apart without cutting it.

Sprinkle over the parsley and let it rest for 15 minutes before serving.

One of the most frequent features of the Greek table is stuffed pasta, which is commonly served as a one-dish meal and I believe you can find a myriad variations all the way up the Central Asian countries, from where it originated. During the Byzantine era, Greeks were introducing several variations of different fillings – especially the Pontians, ethnic Greeks who settled around the Black Sea coast, and whose favourite shape was the triangle.

Pasta triangles

Serves 4

250g semolina or '00' (doppio zero Italian pasta) type flour
2 large eggs
Pinch of salt
50ml whole milk at room temperature
2 tablespoons olive oil
100g manouri cheese
100g feta cheese
50g yoghurt
50g plump moist ready-to-eat sun-dried tomatoes, finely chopped
50g pine nuts
½ bunch of mint, finely shredded
1 litre chicken stock or water
Extra virgin olive oil for drizzling
Salt and freshly ground black pepper
Some kefalotiri cheese to serve

Sieve the flour on to a clean dry surface. Make a well in the middle, crack the eggs into it, add the milk and oil, sprinkle in some salt and bind everything together, preferably by using your hands as the quantity is quite small. Scrape off and clean the surface on which you made the dough, dust it with flour and start kneading the dough until it is springy but quite firm. Let it rest for an hour before you start rolling, which will make the dough softer.

Grate the manouri and crumble the feta into the bowl of a liquidizer or food processor. Add the yoghurt and let the engine run until you get a fine cheesy paste. Turn it out into a mixing bowl and fold in the tomatoes, pine nuts and mint. Mix evenly and set aside.

Divide the rested dough into 4 balls and roll out each ball until they are paper-thin sheets. Put them on a tray that has been sprinkled with semolina. With pastry cutters, cut the dough into discs about 8cm in diameter. Place a small spoon of the cheese stuffing in the centre. Wet one edge of the circle with water on your finger and fold the edge over to make a triangle. Make sure that the edges are sealed tightly.

In a large pot, bring the chicken stock to the boil and drop in the pasta. When the pasta floats to the top, it is ready. Pull the triangles out of the stock with a slotted spoon and place them on plates.

Drizzle with olive oil, season and grate some kefalotiri cheese over them.

8

After the Big Waves, We Are in Every Sense Very Hungry

The local cockerels crow three times a day. The first when the sun takes its early bath on the blue horizon, the second when the sun is at its height and the last when it sinks in the west. This last performance shakes the houses and stirs mankind from being indoors as the smells of food and the lull of music fill the cooler dusk air.

Think of celebrations and communal food where the idea of real hospitality is that everyone dips in the same dish. Bread, olives, fish, sacrificed animals, some figs, goats'-milk cheese – all are eaten and plenty of wine is drunk. Strong onions are reduced to jam softness with a touch of honey and crushed rusks.

Barbecued sea bream with lemon thyme ouzo dressing

The island of Agathonisi is hidden in one of the North East Aegean voids. We entered the narrow strait from the south – the 'little town' and the 'big town' watching down the entrance to the tiny port. With a population of 99 people, there was one school, a power plant combined with a slaughterhouse, a travelling dentist and hundreds of free-roaming goats and young oxen all over the place.

Having been driven off our intended course by very strong winds, we were looking for provisions, and the local 'godfather', a sprightly octogenarian, who ran the harbour shop and taverna, not only arranged for us to be supplied with one of their goats, but sold us a magnificent monster sea bream that had been landed that morning, barbecued it at the roadside and his wife served it up to us for a glorious lunch on the taverna terrace.

Serves 12–16

1 large sea bream, about 7kg

For the stuffing

2–3 celery stalks, including the
*   leaves*
Bunch of spring onions
2 large tomatoes
Handful of oregano
Good bunch of flat-leaf parsley
Good bunch of dill
4 tablespoons extra virgin olive oil
Touch of mastic, pounded
Sea salt and freshly ground black pepper

For the lemon thyme ouzo dressing
200ml extra virgin olive oil
50ml lemon juice
30ml ouzo
1 tablespoon good mustard
1/2 bunch of lemon thyme, picked

Prepare your barbecue to that perfect glowing moment and position the greased grill 15cm above the white coals.

Make the stuffing: coarsely chop all the vegetables (except the tomatoes, which just need to be skinned, deseeded and coarsely diced) and herbs, and mix in a large bowl with all the other ingredients, reserving half the oil and a quarter of the oregano.

Open the fish and clean out the guts, then rinse well inside and out. Fill the stomach cavity with the stuffing. Use wooden skewers to close and secure the cavity. Massage the flesh with a mixture of the reserved olive oil and oregano, and some sea salt.

Place the fish on the barbecue and cook until is done but the flesh is still nice and moist. If the skin starts to burn and/or flake off, try to keep it in place as it will keep the fish flesh moist as it cooks.

While the fish is cooking, make the dressing by stirring all ingredients together until they are well amalgamated.

Serve the fish with the lemon thyme and ouzo dressing. Nice fat chips make an excellent accompaniment.

Warm salad of red mullet, lentils, egg and parsley

Serves 4–5

*150–200ml olive oil or good
    vegetable oil*
6 sage leaves
*1kg red mullet, filleted and pin-
    bones removed*
300g plain flour

For the lentils

250g lentils
2 red onions, grated
4 garlic cloves, peeled but left whole
6 bay leaves
Sprig of rosemary
Salt and freshly ground black pepper
80ml extra virgin olive oil
3 tablespoons red wine vinegar
*100g plump moist ready-to-eat sun-
    dried tomatoes, coarsely diced*
*1 bunch of spring onions, trimmed,
    and cut into thin rounds*
6 soft-boiled eggs, shelled
*1 bunch of finely chopped flat-leaf
    parsley*

First prepare the lentils: rinse and drain them. Place them in a large pot, add the onions and garlic, with the bay leaves and rosemary tied together, then pour in about 1½ litres water (the lentils should be a finger length submerged) and bring to the boil. Turn the heat down and simmer over gentle heat, uncovered, for about 45 minutes, until the lentils are tender. Do not add any salt during cooking or the lentils will harden, and always have the kettle with some boiling water available, just in case they begin to dry out. When the lentils are just tender, switch the heat off.

At this point, season well, pour in the oil, splash with the vinegar, add the tomatoes and spring onions, and mix well. Check the seasoning again (lentils improve a lot with a bit of extra salt) and turn them into a serving bowl, halve the eggs and snuggle them into the contents of the bowl. Sprinkle over the parsley.

Heat 150ml of the oil (use the rest if you have to cook the fish in batches or the pan gets too dry when you turn the filets over) gently with the sage leaves in a frying pan, dust the fish fillets in the flour, shake them to lose any excess and slide them, skin side first, into the oil. Cook for a couple of minutes and then turn and do the same with the other side. Lift the fish out, rest it on kitchen towel or paper for a couple of minutes to drain and then transfer them to the bowl with the lentils.

Bulgur wheat salad with seasonal fruits and cinnamon

Serves 4

200g bulgur wheat (the coarse version)
About 150ml boiling water
Juice of 2 lemons
Juice of 2 oranges
½ cinnamon stick
½ teaspoon cumin seeds, freshly ground
150ml extra virgin olive oil
200g plump moist ready-to-eat sun-dried tomatoes, coarsely chopped
50g fat sultanas
2 pomegranates (or any other seasonal fruit)
1 bunch of spring onions, finely chopped
About 50g flat-leaf parsley, finely chopped
About 50g mint, finely chopped
Sea salt and freshly ground black pepper

Put the bulgur wheat into a bowl, pour over the boiling water, stir and leave for 15 minutes.

Add the citrus juices, olive oil, cinnamon and cumin, and then give the wheat some more time to absorb the new flavourings and become more tender. This can take up to half an hour, depending on the room temperature (the warmer it is, the quicker it will be ready) and on the quality of the wheat. So keep on hand some extra hot water or orange juice.

Add the remaining ingredients (I use a fork to break up the bulgur and mix the ingredients together), season and serve with plenty of warm flat bread, any soft cheeses, yoghurt or even any cold meat leftovers.

The best cheeses that Greece produces are eaten fresh, near where they are made and, as the winter approaches, they are treasured by salting them. The little round pyramid cheese opposite is called touloumotiri and is a kind of semi-soft sheep's curd cheese, which is prepared in sheep's hide, and we used it instead of xinomizithra.

Tirokeftedes

Serves about 12

500g red onions, unpeeled
250g kefalotiri cheese (if unavailable, use pecorino)
250g metsovone cheese (if unavailable, use provolone)
250g xinomizithra cheese (if unavailable, use 125g sour cottage cheese mixed with 125g mature ricotta)
250g feta cheese
1/2 bunch of flat-leaf parsley, finely chopped
1 bunch of mint, finely shredded
Grated zest of 2 lemons
Plain flour, for dusting
4 eggs
Fine semolina, for coating
Good-quality vegetable oil for deep-frying

A few hours ahead, roast the onions whole in an oven preheated to 180°C/gas 4 until the insides are so soft that you can squeeze them out, about 40 minutes. Discard the skin and chop the red onions very finely.

Coarsely grate the kefalotiri and metsovone cheese. Put them in a large mixing bowl, add the xinomizithra, crumble in the feta with your fingers and mix well.

Add the parsley, mint and lemon zest to the mixing bowl and mix well. The finished product should be a coarse stiff paste. You don't need to season! Refrigerate for at least one hour.

To make the tirokeftedes, roll the mixture into small walnut-sized balls 15–20g in weight. Then prepare three bowls: one with plenty of plain flour, another with the well-beaten eggs and a third with plenty of fine semolina. Roll the balls first in flour, dredge in the beaten egg, then roll well in semolina and return them to the fridge for up to an hour. The colder they are, the better they will fry.

Heat the oil for deep-frying to 180°C, then drop the cheese balls into the hot oil in small batches. While they are cooking, begin coating the next batch, and when the ones that are frying have turned golden brown, drain them on absorbent paper.

Bearing in mind that an average person will eat 4, then you have more than enough for at least a dozen of people. They keep well in the fridge, but covered.

Anchovy drippings from the pantry cupboard of Lesvos

The original motive for the production of this sauce was probably to preserve the vast quantities of small fish that were too small for individual consumption but too good to be wasted. Greeks developed a palate for this sauce, garos, and used it as a salt substitute and to add a distinctive flavour to food. It can be as ever-present as ketchup is in this country today.

Makes 750ml

750ml bottle of extra virgin olive oil
10 salted anchovy fillets, well rinsed
1 fresh chilli pepper, bruised
Peeled rind from 3 medium oranges, preferably Seville and unwaxed, shredded

Remove 4 tablespoons of the olive oil from the bottle (use next time you cook).

Chop 5 of the anchovies and put in the bottle of olive oil. Add the whole anchovies, bruised chilli and orange rind. Reseal and let the bottle stand in a dark place for a week or so, shaking from time to time. It will keep for up to 6 months, if stored in a cool dark place.

Use it over salads, grilled fish or simply as a submarine base for your warm bread.

Savoury loukoumades

Loukoumades are the Greek version of beignets, drizzled with honey and dusted with honey. They originate from the Mediterranean Sephardic Jews, particularly those of Spain and Greece. This version is predominantly savoury rather than sweet and can be eaten all year around, but we ate ours in mid-October in the Lilliputian island of Schinoussa. Charming, unspoilt and dreamy are just some of the epithets showering this tiny island in the eastern Cyclades.

Serves 6–8

40g cracked wheat
80ml orange juice with bits
Grated zest of 1 lemon
4 tablespoons olive oil
500g butternut squash, quartered, deseeded and peeled
1/2 bunch of lemon thyme, picked
Salt and freshly ground black pepper
1/2 bunch of mint, finely shredded
1/2 bunch of spring onions, finely chopped
Vegetable oil for deep-frying
A few drops of thyme honey
Pinch of ground cinnamon
Handful of sesame seeds
Feta cheese at room temperature or good yoghurt, to serve

For the batter

10g yeast
125ml water at 25°C
160g plain flour
45ml milk at 25°C
1 medium-sized egg

Make the batter: whisk the yeast into the water. Sift the flour into a large bowl, make a well in the centre and add the yeast water, milk and egg. Mix well to a very smooth thin paste. Cover and leave in a warm place for about 1¹/₂ hours until it rises and bubbles.

Preheat the oven to 160°C/gas 3. In an ovenproof saucepan, mix the wheat with the orange juice, lemon zest and 1 tablespoon of the olive oil. Cover and bake in the oven until the wheat has absorbed all the liquid but it is still wet, about 20–30 minutes.

Turn the oven to 200°C/gas 6. Place the pieces of squash on a roasting tray, brush with olive oil and sprinkle over the thyme and some salt. Roast until soft, about 30–40 minutes. Remove and, when cool enough to handle, chop finely. In a large bowl, mix together the wheat, squash, herbs and remaining olive oil. Do not over-season, as we need to taste the sweetness of the squash.

To make the loukoumades, when the batter is ready, fold it slowly into the wheat and squash mix. Heat oil for deep-frying until very hot but not smoking, about 180°C. Have ready a bowl of cold water to wet a teaspoon so the dough will not stick on it or on your fingers. Take a teaspoon of the dough on the wet spoon and, carefully using your wet fingers, push it off the spoon so it drops into the hot oil. Within seconds it will puff up and rise to the surface. Repeat for about 6 loukoumades at a time, turning them so they become golden all over; this takes only 1 minute.

Take them out with a slotted spoon and drain on kitchen paper. Drizzle a few drops of honey over them, then some cinnamon and the sesame seeds. Eat straight away, with some good feta at room temperature, or dipped in yoghurt, or served with fish.

Pan-seared fig, manouri and pastrouma sandwich

Makes 5

300g manouri cheese
100g feta cheese
2 tablespoons thick Greek yoghurt
Freshly ground black pepper
8–10 figs
10 slices of sourdough bread
  (or any other good bread)
A few rocket leaves
250g sliced pastrouma (see page
  220 or, if unavailable, use
  pastrami)
2 teaspoons aged red wine vinegar
  or balsamic vinegar
Butter, preferably clarified
  (see page 186), or ghee, for
  frying

On a chopping board, mash the cheeses with a knife or fork. Add the yoghurt and pepper, and mix well. (Needless to say, you can do the same in a blender.)

Cut off the end of the figs and, without removing the skin, cut them lengthwise into 2–3 slices.

Spread the cheese mixture on all the bread slices. On that base, build up layers of rocket, pastrouma and figs on half of the slices. Splash the tops with a few drops of the vinegar and cover each with another piece of bread.

Put some butter in a large frying pan and place over medium-low heat until foamy. Add the sandwiches, pressing them into the pan, and fry slowly, regulating the heat so the butter does not burn. Once light brown, turn them over, and press down with a spatula to compress slightly. Brown the other side and serve straight away.

Potting meat at the end of October

Meat potting is preserving meat in a large container. This is how we did it last year when we visited Syros. Early in the morning one of the locals killed a pig and started cutting it up. He gave the pieces to me and I started frying the pork and filling up a crock-pot with it.

Serves 8 with leftovers

250g butter
Several fig or vine leaves (optional)
2kg stewing meat, beef, veal, lamb
    or pork
10 bay leaves
100g plump moist ready-to-eat sun-
    dried tomatoes
2 cinnamon sticks
2 cloves
1 fennel bulb, trimmed and
    quartered
1 small head of garlic, sliced in half
Salt and freshly ground black
    pepper
1 bottle of red wine

Preheat the oven to 130°C/gas 1. Preferably use a large straight-sided earthenware pot, failing that a heavy-based straight-sided casserole, grease it thoroughly with some of the butter and if you're by a fig tree use some leaves to line the bottom, otherwise use vine leaves, or nothing!

Cut the meat into 3cm-thick slices, trimming off and discarding any gristle and sinews. A good butcher can do this job for you.

Put the bay leaves, tomatoes, cinnamon and cloves on the bottom of the pot. Lay in the meat slices, then on top of that the fennel and garlic. Season. Pour in the wine. It should just cover the meat.

Simmer the pot, uncovered, in the oven for 1½–2 hours, until the meat is very tender and the volume halved. Remove and let cool.

Drain off the liquid and set aside. Discard the cinnamon and cloves. Pop the garlic out of its skin and finely chop it with the meat and fennel. Adjust the seasoning. Return the meat, garlic and fennel to the pot, pour over the reserved juices and shake well.

Clarify the remaining butter by warming it gently in a small pan so its solids fall to the bottom of the pan, skim off the foam and then carefully pour off the clarified butter, leaving the solids behind. Seal the potted meat with a layer of this. Cover with wax paper and tie down with string.

Leave the pot refrigerated, where it will keep for several weeks, as long as you always seal off with more clarified butter the part you've removed. You will now have a delicious dark-jellied meat, which you can eat either hot or cold.

Goat on grilled pita with purslane and grated cheese

If you can't get goat, then this is equally good made with lamb – or, even better, mutton.

Serves 6–8

120ml olive oil

6 bay leaves

1 shoulder of goat, boned and cut into 3cm dice

300g leeks, trimmed, well rinsed and finely chopped

6 garlic cloves, peeled and left whole

100g plump moist ready-to-eat sun-dried tomatoes

150ml red wine

Salt and freshly ground black pepper

1 teaspoon red pepper flakes

300g purslane, thick stems discarded and leaves coarsely shredded

1 bunch of flat-leaf parsley, finely chopped

3 tablespoons fresh lemon juice

6–8 flat breads

A chunk of kefalotiri (if unavailable, use pecorino), for grating

Sliced pickled cucumber, to serve

Extra-thick yoghurt, to serve (optional)

In a large straight-sided casserole, heat the oil gently with the bay leaves. Add the goat and cook over moderate heat, stirring occasionally, until all pieces are browned evenly, about 5–8 minutes.

Stir in the leeks and garlic. Cover and cook for a few minutes until both vegetables are softened but not browned. Add the tomatoes and wine, and bring to a simmer. Season with salt and red pepper flakes, cover and simmer until the mixture begins to caramelize and the meat becomes very tender, about 45 minutes.

Remove from the heat, strain the liquid and return it to the casserole. Turn the meat on to a large chopping board, chop it as finely as you can and put it back in the casserole. Stir in the remaining oil, purslane, parsley and lemon juice, mix well, adjust the salt to taste and let it stand for up to 30 minutes.

Spread the mixture over grilled flat bread with some grated kefalotiri cheese and sliced pickled cucumber. I sometimes also add a dollop of extra-thick yoghurt.

Grilled artichokes and lamb tongues

Serves 6–8
For the lamb tongues

1kg fresh lamb tongues
12 black peppercorns
1 tablespoon salt
6 bay leaves
1 tablespoon fennel seeds
2 lemons, sliced
1 onion, cut in half

For the artichokes

6 medium-sized artichokes
4 lemons
300ml extra virgin olive oil
1 tablespoon aged red wine vinegar
  (if unavailable, use balsamic)
3 garlic cloves, peeled and crushed
Freshly ground black pepper
1 bunch of flat-leaf parsley
1/2 bunch of mint, picked and
  shredded at the last moment
100g natural green cracked olives
50g capers

To serve

soft-boiled eggs, shelled
breakfast radishes
pickled cucumbers

You need to start preparing the tongues several hours ahead. Put them in a large pot with the peppercorns, salt, bay leaves, fennel seeds, lemon slices and onion halves. Cover with cold water, bring to the boil and simmer for about 2 hours, until the tongues are tender. Remove them from the stock and let them cool sufficiently to be able to handle them.

The skin of the tongue's taste-buds side will have pulled away a bit from the meat. Slit this with a small sharp knife, but don't cut the flesh, then just peel off both sides and cut the root end off.

Prepare the artichokes: fill a large bowl with cold water and add the juice of 2 of the lemons. Cut off the stem and top quarter of each artichoke. Peel off the dark green outer leaves and snap off the base until only the pale green and yellow leaves remain. Trim any dark green areas off the base. Cut each artichoke in half lengthwise, then scrape out and discard the fuzzy centre and any purple-tipped petals. Drop the halved artichokes into the lemony water.

Prepare the barbecue.

Steam the artichokes until almost tender, about 20 minutes. When they're ready, mix 100ml olive oil, the aged red wine vinegar and the chopped garlic in small bowl and brush this over the artichokes. Sprinkle the artichokes with salt and pepper. Grill the artichokes on the barbecue until slightly charred, turning them occasionally, about 8 minutes.

Pour the remaining olive oil into a large mixing bowl and whisk in the juice from the remaining lemons. Add the herbs, olives and capers, and stir once more. As the artichokes come off the grill, transfer them into a serving bowl and pour over the dressing.

Brush the tongues with the remaining garlic oil from the artichokes (top it up if it is not enough) and grill them for a few minutes on each side. Add them to the artichokes and eat while warm, with soft-boiled eggs, breakfast radishes and pickled cucumbers.

Calves' liver with celeriac and avgolemono sauce

If you're feeling a little under the weather, nothing will perk you up like a nice bowl of celeriac avgolemono.

Serves 4 and a child

500g calves' liver
100g plain flour
50g fine semolina
*1 teaspoon fennel seeds, coarsely
    ground*
35g butter
150ml olive oil

**For the celeriac and
avgolemono sauce**

25g butter
50ml olive oil
*About 750g celeriac, peeled and cut
    into walnut-sized pieces*
*200g leeks, trimmed, well rinsed
    and coarsely chopped*
300ml chicken stock
About 50ml lemon juice
2 eggs
Salt and freshly ground black pepper
*At least 1 bunch of dill, finely
    chopped*

First start the celeriac and avgolemono sauce: in a heavy-based straight-sided casserole, very gently heat the butter with the olive oil. Add the vegetables and sauté for a few minutes. Pour in the chicken stock, cover and simmer for 15 minutes, until the celeriac is tender. How tender? About as tender as you like them.

Slice the liver into 4 portions, remove any sinew as you cut and keep in mind that you need to end up with a thickness of 2cm.

Finish the sauce: beat the lemon juice and eggs together until light and frothy. Slowly beat a ladleful of the warm sauce into the egg mixture; whisk vigorously, and then stir it back into the pot. Stir well to make sure the eggs are amalgamated, so that they won't curdle. Adjust the seasoning, add the dill and keep it warm, preferably in a bain-marie, otherwise over very low heat.

Mix the flour and semolina in a shallow bowl, season it with salt and fennel seeds and drag the liver into it. Heat the butter and olive oil gently in a large frying pan. When the butter starts foaming, slide the liver into the hot oil and fry each side for a minute, until seared, lightly browned and with a crisp topping. I eat my liver juicy, thus pink inside.

Serve the liver with the sauce.

Salad of green cracked olives, pistachios, egg and sesame seeds

The hot dry climate of Greece creates perfect growing conditions for olives and citrus trees, which both play an important part in simple basic Greek cooking.

Serves 6–8

400g green cracked olives

2 medium-sized red onions, very thinly sliced with a mandolin

100g breakfast radishes but you can use other ones too (shred the leaves into the bowl if they still look fresh)

1 bunch of mint, finely shredded

1/2 bunch of lemon thyme, picked

1/2 teaspoon fennel seeds, roughly ground

80ml extra virgin olive oil

1 orange, peeled and separated into segments

100g pistachios, preferably from Aigina, roughly crushed

1 tablespoon sesame seeds

8 semi-soft boiled eggs, shelled

Drain the olives, rinse them several times under cold running water and place them in a large mixing bowl. Add all the remaining ingredients, apart from the sesame seeds and eggs, toss well and set aside to let the different flavours and textures feel less 'foreign' with each other!

Half an hour later, cut the eggs in half and drop them into the salad, then sprinkle the sesame seeds over.

9

Sugar and Honey Make the Storm Lantern in our Hearts

We all know that sugar and honey swim on the lookout for the ideal marriage.

Their course has no end; during the day they craftily sail the local kitchens, and by night they make

their way back in our dreams. Their perfect partnership makes a powerful catalyst to bring out

the best in the Islands' fruit and grains, producing a wealth of wonderful dishes.

After-lunch Aegean mint fruit salad

Serves 6–8

1 small melon
3 peaches
3 nectarines
300g apricots
300g seedless white or black grapes
150g thyme honey
*200ml Muscat wine or any good
    dessert wine*
1 bunch of mint leaves, picked
*Manouri cheese and lots of freshly
    ground black pepper, to serve
    (optional)*

**For the caramelized
walnuts**

60g brown sugar
50g butter
60g thyme honey
*2 teaspoons aged red wine vinegar
    or balsamic vinegar*
150g walnuts

First prepare the caramelized walnuts (ideally the day before): preheat the oven to 160°C/gas 3 and line a heavy rimmed baking sheet with baking parchment.

Combine the sugar and butter in saucepan, and stir over medium heat until the sugar dissolves, about 3 minutes. Add the thyme honey, vinegar and walnuts, shaking the saucepan to coat them.

Transfer the mixture to the prepared baking sheet and bake until the nuts are deep brown and the syrup thickens and coats the walnuts, stirring occasionally, about 30–40 minutes. Allow to cool completely on the baking sheet.

Separate the walnuts by breaking them up, and store in an airtight container at room temperature.

Cut the melon into wedges, remove the seeds and skin, and cut the flesh into small bite-sized pieces. Do the same with the peaches, nectarines and apricots. In a large salad bowl, mix the melon with the rest of the cut fruit and the grapes.

Over very low heat, dissolve the thyme honey in the Muscat wine, mix well, remove from the heat and let cool completely.

Pour the mixture over the fruits and refrigerate for at least an hour. Before serving the fruit salad, add the caramelized walnuts and mint leaves. I like mine with some manouri cheese and lots of freshly ground black pepper.

The last week in August is when the local markets are inundated with the subtly crunchy and thirst-quenching watermelon, which grows everywhere. Greeks love them when they are perfectly ripe, and so sweet and lushly juicy that they make the sugar disappear from our hearts. What else can bring the meal to such a sweet and luscious end?

Watermelon and other fruits for grown-ups and kids

Sultanina is a variety of tiny ellipsoidal seedless grapes, which when they reach the zenith of their ripening have a tempting sweet, tasty pulp. The grape ripens from about the end of August to the beginning of September, and Crete has probably the best climatic and soil conditions for sultanina plantations.

We approached them with the dusk to collect a few for our dessert, and from far away in the September light they looked like shiny pebbles, lonely, deserted, mysterious, but they had everything the palate could desire. Paloma, I beg your pardon for losing Freddie in the fields. And because of that I wasn't able to make this recipe on board, but here is the London reconstruction!

Serves 6, plus some for the next day too!

1 watermelon, about 4kg

300g peaches, peeled and coarsely grated

200g apricots, stoned and finely chopped

300g grapes, preferably sultanina

50g fat raisins

100g thyme honey

120ml Muscat wine

Handful of fresh mint leaves

Several hours ahead, remove the rind and seeds from watermelon and cut the flesh into chunks.

In the bowl of a food processor, purée the watermelon until completely liquefied. Transfer it to a large serving bowl, stir in fruits, raisins, honey, wine and mint.

Cover and refrigerate for several hours.

Strawberry pelte

Pelte is a strained fruit jelly that is very common on the Ionian Islands and was the most popular preserve of royal Ottoman cuisine.

Makes about 400ml

1 kg strawberries
Granulated sugar (see below)
Juice of 1–2 lemons

Remove the stems from the strawberries and slice them into a heavy straight-sided pan. Add 250ml water, bring to the boil and boil for 30 minutes. Add another 100ml water, bring it to the boil and then simmer gently for about 1 hour.

Take a large cheesecloth and drape it over a deep bowl. Pour the strawberries and liquid into the cloth, tie the cloth and suspend it over the bowl so the juice drips into the bowl.

About 24 hours later, measure the juice and, for each cup of juice, add in half a cup of sugar. Put back in the casserole and stir over medium heat. Add the lemon juice and boil for about 30 minutes, skimming frequently.

Test a teaspoonful of pelte on a chilled plate; it should almost set, otherwise put it back over the heat. Leave to cool.

When ready, pour it into one or two warmed sterilized jars and seal them. They will keep, unopened, for at least 6 months.

Delightful September's figs –
fig saganaki

The Greek Islands are a fruit lover's paradise. There are fruit trees in every garden; and along the small village streets as well, so you can often help yourself.

Serves 6–8

1kg not-too-ripe figs
50g butter
1 tablespoon extra virgin olive oil
300g manouri cheese
3–4 tablespoons thyme honey
¹/₂ teaspoon freshly ground
    cinnamon
Freshly ground black pepper
1 bunch of fresh mint, finely
    shredded
2 tablespoons Corinthian aged red
    wine vinegar (optional)

Preheat the oven to 200°C/gas 6. Trim away the stem at the top of each fig and cut each fig in half lengthways.

Melt the butter with the oil in a flat gratin dish over medium heat, place the figs in it, cut side up. Slice the manouri cheese and lay it over the figs. Drizzle over the honey, sprinkle with the cinnamon and pepper, and roast for about 10–15 minutes until the figs are slightly softened. The juice from the figs and the honey should make good syrup.

Sprinkle over the mint and allow to cool slightly in the dish. A few drops of vinegar will lift the flavours, but this is optional.

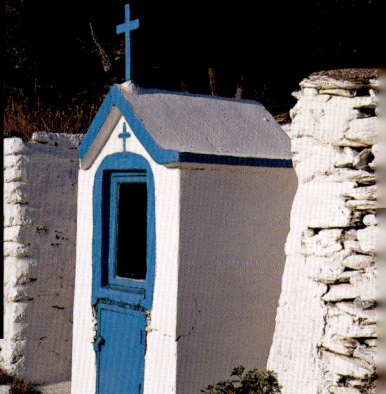

Grilled trahana and peach with yoghurt

Cooked, cooled and set, the trahana is brilliant grilled until crisp and golden.

Serves 6 with leftovers

200g granulated sugar
150ml Muscat wine or other good dessert wine
2 vanilla pods, split lengthwise and seeds scraped out
1kg peaches, peeled, cored and sliced into 2cm-thick wedges
750ml unsweetened grape juice
200g trahana (see pages 14–15)
100g butter
2 eggs
Thick yoghurt, vanilla ice cream, or feta at room temperature, to serve

In a large saucepan, combine the sugar, wine, vanilla bean and its seeds with 150ml water. Set over low-to-medium heat, bring to a simmer and cook, stirring occasionally, until slightly syrupy, about 10 minutes. Remove from the heat, stir in the peach wedges and allow to cool to room temperature.

Remove the vanilla bean pieces from the syrup, with a slotted spoon. Remove a bit more than half of the peaches and place the remaining fruit and syrup in a food processor. Purée these peaches until smooth.

Preheat the oven to 200°C/gas 6. In a large saucepan, bring the grape juice and fruit purée to the boil. Have a long-handled wooden spoon ready in one hand and pour the trahana in a steady stream into the liquid, stirring all the time. Reduce the heat and continue to stir for about 10 minutes until the trahana becomes thick and smooth. Stir in the butter, minus a knob that you will need for brushing the baking dish. Remove from the heat, let cool for 10 minutes and stir in the eggs.

Brush with the reserved butter a heavy-based shallow baking dish big enough to take all the ingredients. Spread the trahana out to form a slab 2 cm thick and then snuggle in the reserved peaches, forming a military parade in a line. Bake for 20–30 minutes in the preheated oven. Allow to cool completely.

Preheat a medium grill. Cut the baked trahana into wedges, about 8x4cm. Grill the wedges for about 3 minutes on each side, until charred and golden.

Try it with creamy, thick, stand-your-spoon-up-in-it yoghurt, or vanilla ice cream, or just room temperature feta. It is sheer bliss.

Sfakianopites (cheese pies from Sfakia, Crete)

Just-made mizithra cheese melts on the tongue and fills the palate with fathomless sweet warmth. You find mizithra in two different varieties – fresh and aged. Fresh mizithra is unsalted and soft, with a pungent aroma and mild flavour. Similar to ricotta, it is often sold in egg-shaped balls and preferably should be eaten on the day it is bought. Aged mizithra is a hard salty cheese, used for grating over pasta dishes, soups and vegetable casseroles.

Serves 6 (makes about 20)

500g 'OO' (Italian doppio zero pasta) flour
2 tablespoons ouzo
4 tablespoons olive oil
Pinch of salt
500g fresh mizithra cheese
1 egg, lightly beaten
30g butter
A few tablespoons thyme honey
1 tablespoon ground cinnamon

Sift the flour on to a worktop, make a well in the middle and slowly pour in about 300ml water, the ouzo, 2 tablespoons of the oil and the salt. Knead to make a soft dough. Cover with a damp cloth and let it rest for one hour.

Roll it out to a thickness of 3mm at the most. Using a pastry cutter, cut as many 10cm rounds out of the dough as you can, re-rolling the trimmings. Roll the cheese into ping-pong-sized balls. Drop one into the centre of each of half the dough rounds and flatten slightly. Brush the edges of each of those rounds with egg and place another round over them to cover the cheese and seal by pinching the edges together.

Glaze the bottom of a medium hot frying pan with the remaining oil and the butter, and fry the pies until browned on both sides.

Serve hot, drizzled with honey and sprinkled with cinnamon.

Green tomato ice cream

Makes about 1.5 litres
450ml full-fat milk
550ml double cream
1 vanilla pod
7 egg yolks
175g sugar

**For the preserved
tomatoes**
50g sultanas, ideally Corinthian
100ml Mavrodaphni or good port
Peeled zest of 1 lemon
150g sugar
*2 cinnamon sticks, each broken into
    couple of bits*
½ teaspoon black peppercorns
3 cloves
*500g green tomatoes, stalks removed
    and cut into 2cm wide wedges*
*100g pistachios, shelled and roughly
    crushed*

At least a couple of days ahead, start preparing the preserved tomatoes: soak the sultanas in the Mavrodaphni, preferably overnight.

In a medium saucepan, mix the lemon zest, sugar, cinnamon, peppercorns and cloves with 300ml water and bring to boil. Cook for 5 minutes.

Add the tomatoes and sultanas with their soaking liquid, bring back to the boil, reduce the heat and simmer, stirring frequently, until the tomatoes are soft but still hold some of their shape, about 20–30 minutes.

With a perforated spoon, transfer the tomatoes to a bowl, leaving the other ingredients simmering until the syrup can coat lightly the back of a spoon. If you have a sugar thermometer, it will show about 110°C.

Strain it over the tomatoes, add the pistachios and let it cool. Transfer to sterilized jars and seal. This will now keep, refrigerated, for several months.

Make the ice cream at least the day before you need it: in a heavy-based saucepan mix the milk and cream. Slit the vanilla pod in half, scrape the seeds into the milk mix and bring gently to a simmer (no more than 80°C). Turn off the heat.

In a large bowl, whisk the egg yolks with the sugar for a few minutes. Pour in the hot milk, whisking continuously to avoid scrambling the eggs. Return the eggy sweet milk back to the saucepan and cook over very low heat, stirring with a wooden spoon, until it turns into a cream that coats the back of the spoon.

Pass it through a sieve into a container and cool down quickly, preferably in a very cold bain-marie. Refrigerate.

The next day, churn it in an ice-cream machine. When the ice cream is almost frozen, add 300g of the green tomatoes.

Manouri and pistachio ice cream

Makes about 1.5 litres

250g roasted pistachios
*300g manouri cheese, grated (if
    unavailable, use mascarpone)*
400ml double cream
600ml full-fat milk
100g sugar
6 egg yolks
*Dash of freshly ground pepper
    (optional)*
75g thyme honey

Make this at least a day ahead: put the nuts in a food processor and whiz them up until you get a smooth paste.

Put the manouri cheese in a food processor with the cream and milk and whiz up. Transfer the blended mixture to a saucepan over a medium heat.

In a large mixing bowl, whisk together sugar, egg yolks and pepper, if you are using it, until fluffy and pale yellow.

Slowly blend the hot milk mixture into the egg mixture, one ladleful at a time, stirring constantly to prevent the eggs from scrambling. Return to the saucepan, add the pistachio paste and cook over a very low heat, stirring continuously with a wooden spoon, until it turns into a cream that coats the back of the spoon.

Remove from the heat and cool down quickly, preferably in a very cold bain-marie. Refrigerate overnight.

The next day, churn it in an ice-cream machine and, when the mixture is almost frozen, fold in the honey. Pour it into a container and keep it in the freezer until it sets.

Pasteli

A great source of energy, pasteli is a speciality of Kea, a mountainous Cycladic island a few miles away from Attica, which is also known for its production of thyme honey. The process of making pasteli is very simple.

Makes about 50 pieces

450g thyme honey

450g sesame seeds, toasted in the oven

2 slices of mandarin zest, finely chopped (optional)

In a heavy-based pot, bring the honey and 100ml water to the boil. If by any chance you own a sugar thermometer, let the temperature reach the 130°C; otherwise, when it starts firming up, remove it from the heat. Stir in the sesame seeds and mandarin zest, if you are using it.

Cover a clean flat surface with baking parchment, spray this lightly with water and spread the mixture with a spatula evenly over it to a thickness of about 1cm. When it has hardened, cut it into rectangular pieces.

Layered between sheets of baking parchment in an airtight container in a cool dry place, it will keep for a bit more than a week.

Frozen Ouzo Mojito

Ouzo in mojitos is down to one of London's most famous club barmen, Nick Strangeway. If you haven't the time or inclination to make a syrup, a bottle of gum syrup will last for many rounds.

Makes 2 glasses

24 mint leaves
100ml lime juice
Crushed ice
100ml ouzo
Soda water to taste

For the mint syrup

150g granulated sugar
50g brown sugar
24 mint leaves

Make the mint syrup: over medium heat, dissolve the sugars in 300ml water. Then infuse the mint leaves in the syrup for about 1 hour. Keep it in your fridge until is needed or finished.

To make the mojitos: divide the mint leaves and lime juice between 2 glasses and gently bruise the leaves with a blunt instrument of your choice.

Pour 25ml of the sweet mint syrup into each glass and add some crushed ice, the ouzo and soda water to taste. Stir well and sip through straws.

Platanakia
Πλατανάκια
Aviakia
Αυλάκια
Tsamadou
Τσαμαδού
Lemonakia
Λεμονάκια
Mytilinii
Μυτιληνιοί
Chora
Χώρα
AIRPORT
ΑΕΡΟΔΡΟΜΙΟ
Vathi
Βαθύ
SAMOS
ΣΑΜΟΣ
Nisi
Νησί
Agia Zoni
Mourtia
Μουρτιά
Kerveli
Κερβέλι
Paliokastro
Παλιόκαστρο
Psili Ammos
Ψιλή Άμμος
Mykali
Μυκάλη

A Brief Guide to the
Greek Islands and Their Food

Most of Greece's 1,400 islands are scattered across the Aegean Sea, bordered by the Greek mainland to the west and north, Turkey to the east, and Crete, Greece's largest and most southerly island. Inter-island travel can be a challenge. The most common phenomenon of Greek Island exploration is couple of days of blustery weather that creates havoc in our daily lives but offers some unconditional love as long as your mind can be free of limitations. They are part of me.

The Greek Islands grew as an amalgam of independent city-states, many of which established colonies throughout the Mediterranean. Many later fell sway to various colonizing forces through the centuries, notably the Roman, Venetian, Ottoman and British Empires, as well as the French and Italians. They are commonly divided into groups as below.

Crete, the largest of all the islands by far was the centre of Europe's most ancient civilization, the Minoans. Much later it was ruled by the Venetians and then by the Ottoman Empire until 1897, so is possibly the most Turkish-influenced of all the islands.

The Cyclades lie in the centre of the Aegean between Chios and mainland Greece and down towards Crete. They are the classic Greek islands that first come to mind when we see images of the whitewashed villages. Mykonos, Paros, Naxos, Siros, Ios, Santorini

(presumed to be the legendary Atlantis) and the historic island of Delos, are some of the islands in this group. Fairly untainted by outside influence through the centuries, they are noted for the particularly imaginative ways the islanders make meals from the fairly meagre local produce.

The Dodecanese ('Dodeca' is the Greek word for 'twelve' and there are twelve islands in this group) run along the Turkish coast. My favourite ones are Patmos, Fournoi, Lipsi and Karpathos. These islanders have a very independent spirit and are known for their good traditional – and often quite spicy – food.

Evia is the second largest island after Crete and has been inhabited since prehistoric times. The island's name means 'rich in cattle'. It is often perceived as part of the mainland and is actually joined to it at Attica by a bridge. Food in Evia is generally of a higher standard than is to be found on those islands more popular with tourists as it has long been a favourite haunt of more discerning Greek holiday-makers.

The Ionian Islands lie off the west coast of Greece in the Ionian Sea. They are Keffalonia, Ithaca, Lefkada, Paxoi, Kerkyra (Corfu), Kithyra and Zakinthos. Because of their position and a history of being ruled by the Venetians, the French and the British, they are possibly the most

'European' of the islands in terms of food and culture.

The North Eastern Aegean Islands lie very close to the Turkish mainland and include the wild island of Icaria, Chios, Samothraki, Lesvos and a few more. The islands are quite disparate in their cultures and food, although Lesvos and Chios are the main ouzo-producing islands.

The Saronic Islands are very close to Athens, and include Aegina, Hydra, Poros and Spetses.

The Sporades are a small group of islands to the north-west of Athens in the northern Aegean. Alonisos, Skiathos and Skopelos are some of them. As with Evia, both groups share much of the food and culture of the mainland.

The Greek Island-hopping Calendar
Through Flowers, Festivals, Sounds and Smells

January is the pruner's month. Vines and some trees are pruned, but not on January 1st as it is the name day of Aghios Vassilis (St. Basil), the Greek 'Santa Claus'. That day, we slice the Vassilopita, a rich brioche that was given its name in his honour, and the person who finds the hidden coin in his or her slice is sure of good luck for the year. Also January 6th is the day of Epiphany and the blessing of the sea. A cross is thrown into the sea and retrieved by shivering swimmers. In this month some of the islands are dotted with purple anemones.

February is the month of meat and cheese Sundays – the two Sundays before Lent. During this month the deliciously scented narcissus fills our nostrils, and the high spot of the month is the carnival, which lasts for two weeks.

March is a busy month: the fully blossomed orange and lemon trees in the bay of Vathi in Kalymnos are so beautiful. It also heralds the first spring winds, the first sound of the southern Aegean cicadas and, often Clean Monday (see page 162) falls in March, 40 days before Easter, marking the first day of Lent, with the skies filled with kites.

April, with its magic smell of orchids, usually contains the Greek Orthodox Easter (but not always, so enquire at the local Greek Church for an accurate date). Red eggs (hardboiled with a red dye to represent Christ's blood), Magiritsa (a soup made out of the lamb's entrails), roasted lamb on the spit and wine are all in abundance. My favourite place to be during pre-Easter time is in Hydra; a unique experience with the majestic local custom of immersing the Epitaph (a wooden representation of Christ's tomb) in the sea on Good Friday evening on Kamini beach. Saint George's Day, on April 23rd, is celebrated on Lemnos in the village of Kaliopi with horse races.

May is the beginning of the outdoor living, picnicking and partying. The wild thyme begins to flower, to the delight of the local bees. The month begins with Labour Day and the feast of the flowers.

June sees the rock caper plants start painting the rocks and cliffs with their long purple stamens. Towards the end of June, in the town of Agia Paraskevis in Lesvos, sees a three-day festival of the bull, with live music, celebration and horse races through the streets.

July is marked by most of the Aegean Islands being dressed with prickly pears. Around the end of the month in Sitia, Crete, the sultana and raisin festival takes place.

August is the 'dancing month', with numerous festivals. My favourite is the anchovy festival during the first week of August. Kaloni, a small village in Lesvos holds a great festival with the fish, ouzo, live music and dancing. August 15th is the day of the Virgin Mary, and on the island of Tinos it is celebrated by pilgrims gathering by their thousands to crawl on their knees up the steps to the splendid church built to hold a holy icon found there in the 19th century. It is thought to be the work of the apostle St Luke and buried to save it during the Ottoman invasions.

September and all the eyes pinpoint the grape harvest.

October is distinguished by cyclamens flooding the islands with their lilac colour, and the grape gathering for the sweet wines is in full swing.

November is the month to be in Crete for the combined power of the purple blossoms of the saffron crocus, the first glimpse of snow caps on its mountains and the poignant smell of the cyclamens that still warms the hearts of the island people.

December has Saint Nicholas as its patron saint, and he is also the patron of sailors. Oranges and tangerines overindulge our palates.

Glossary of Greek Ingredients

Aged Corinthian vinegar: an intense dark-coloured vinegar with a delicate fruity flavour and sweet clean aftertaste. It is produced from Corinthian grapes grown in the Kalamata region of the Peloponnesus, and then aged in wooden barrels. See also page 129.

Anthotiro cheese: a variation of Mizithra and is also a sheep's or goats' milk cheese. Like Mizithra, it comes in two versions, soft and ricotta-like and dry and salted, although only the former is used in this book.

Avgotaraho: also known as bottarga, this is salted dried grey mullet roe; see page 90.

Bonito: (*Sarda sarda*) a fish closely related to the tuna and like a smaller version of that fish; see also pages 156-7.

Bourekia: little pies made from filo pastry; see page 119.

Bulgur (pourgouri): also known as bulghur wheat, this grain is made from wheat kernels that have been steamed, dried and crushed. See also page 39.

Elephant bean: a large variety of white kidney bean, see page 122.

Feta cheese: in 2002, Greek feta was awarded EU Protected Designation of Origin (PDO) status, giving Greek producers in specific regions exclusive rights to the 'feta' name.

Feta is salted and cured in a brine solution (which can be either water or whey) for several months. Feta dries out rapidly when removed from the brine. Feta cheese is white and can range from soft to semi-hard, with a tangy, salty flavour that can range from mild to sharp. Its fat content can range from 30 to 60 percent; most is around 45 percent.

The feta we ate on the boat is produced by Roussas, exclusively from sheep's and goats' milk in the mountainous region of Almyros, central Greece, which has become an officially recognized feta-producing province. The herds graze freely on the rich geo-specific flora, which contributes to the flavour of the cheese.

Flomaria (hilopites): the traditional egg noodles of Limnos; see pages 142-3.

Froutalia: a Greek version of the frittata; see page 125.

Graviera Kritis: a hard cheese from Crete with a Protected Designation of Origin (PDO) status. It is a yellow cheese with a hard rind, an intense aroma and a sweet, mellow and slightly nutty flavour.

Greek strained yoghurt: traditional Greek yoghurt produced from sheep's milk, which is turned into yoghurt then strained through cloth until it reaches a rich and creamy texture; see page 148.

Kalamata olives: large black olives, named after the city of Kalamata and used as a table olive, as they have a fine smooth texture and meaty taste.

Kasseri: a Protected Designation of Origin (PDO) cheese made from sheep's and goats' milk. Matured for at least 3 months, it has a mellow, yet slightly piquant, flavour and is yellowish-white in colour.

Kataifi: a shredded filo-like pastry resembling angel hair pasta; see page 65.

Kefalotiri: a hard yellow cheese made from sheep's and goats' milk, with a sharp, salty flavour; see also page 121.

Loukanika: Greek sausages; see page 124.

Loukoumades: the Greek version of the doughnut; see page 181.

Manouri: traditional Greek PDO cheese made by taking the whey from feta production and adding up to 60% cream from sheep's milk. The rich and creamy version of this white cheese we had on board was from the Roussas diaries. See also page 45; if unavailable, use mascarpone.

Mastic: an aromatic resin, used as a flavouring, derived from an evergreen shrub of the pistachio family and mostly cultivated on the island of Chios; see also page 20.

Mavrodaphne: a Greek sweet fortified red wine. If unavailable, use good port

Metsovone: a semi-hard smoked cows'-milk cheese.

Mizithra: another ricotta-like Greek cheese; see page 207.

Okra: see page 42.

Orzo: tiny rice-shaped pasta, often used in soups.

Ouzo: anise-flavoured liqueur widely popular in Greece and similar to French pastis and Turkish raki, but a little sweeter and smoother. It can be taken either as it is or with water; see page 79.

Pastrouma: spiced cured beef, similar to pastrami; see page 185.

Perasti: see Tomato perasti.

Pelte: a fruit jelly; see page 203.

Petimezi: a syrup made from grape juice.

Pinnes: a large bivalve shellfish, also known as the fan mussel; see page 110.

Pourgouri: see Bulgur.

Purslane: (*Portulaca oleracea*) is an annual succulent plant used both as a salad leaf and cooked as a vegetable. Slightly sour and salty, it is very popular throughout Greece.

Santorinian tomato: a variety of cherry tomato native to the island of Santorini that requires no watering and has a wonderful intense flavour; see page 36.

Savoury graviera or kephalograviera: also from Crete, savoury Graviera is matured for longer and is therefore saltier and has a slightly more piquant flavour.

Sfakianopites: cheese pies, see page 207.

Spoon sweets: see page 10.

Sultanina: tiny sweet grapes; see page 201.

Throumbes olives: dry-cured black olives, see page 16.

Tirokeftedes: cheese balls; see page 182.

Tomato perasti: a thick and intensely flavoured Greek tomato passata produced from pressed, strained hand-selected tomatoes; see pages 128-9.

Trahana: a homemade dried pasta; see pages 15, 137, 138 and 206.

Tsipouro or raki: a distilled spirit made from wine press residue.

Vine leaves: hand-selected young, pale-green and very tender leaves from the grape vine are used to roll up savoury fillings to make dolmades. Fresh, they need only the briefest of blanching; bought, they have usually been preserved in brine so need thorough rinsing.

Visanto: a sweet red dessert wine from Santorini.

Yoghurt: see Greek strained yoghurt.

Yoghurt cheese: see page 133.

Xinomizithra: a sheep's or goat's whey cheese from Crete which is yet another Mizithra variation and has an intriguingly sour taste.

Index

A
Aegina 13
Agathonisi 176
anchovies: anchovy drippings 182
  anchovy fillets with Santorinian capers 79
  fried squid with a salad of artichoke, potato and anchovy 101–2
Andros 125
Antikythira 55
Antiparos 105
apples for Freddie 127
apricots: prawns with apricot feta salad 87
  stuffed cuttlefish 108–9
artichokes 100
  fried squid with a salad of artichoke, potato and anchovy 101–2
  grilled artichokes and lamb tongues 188
aubergine stifado 49
avgotaraho (grey mullet roe) 90
  sea urchins, avgotaraho, fennel and kohlrabi salad 91

B
basil oil 70
beans see borlotti beans; broad beans; elephant beans
beef see pastrouma
beetroot with fried marida 45
biscuits, roasted pistachio 12
bonito and Santorinian spring vine leaf rolls 156
borlotti beans with grilled chicken wings and purslane 54
bougatsa (savoury filo pie) 24
bourekia me revithia 119
bourgiourdi (roast feta en papillote) 62
bread: crab meat on grilled sourdough bread 93–4
  goat on grilled pita 187
  pan-seared fig, manouri and pastrouma sandwich 184
  parsley and onion dip 164
broad beans with braised lamb sweetbreads 41
bulgur wheat: bulgur, walnut and spinach pilaf 39
  bulgur wheat salad with seasonal fruits 179
  Cretan dolmades 32–3
  trahana 138
butter, cherry 12
butternut squash: savoury loukoumades 183

C
calves' liver with celeriac and avgolemono sauce 191
celeriac: calves' liver with celeriac and avgolemono sauce 191
  celeriac roots and watercress 159
cheese 180
  beetroot with a manouri and yoghurt dressing 45
  bougatsa (savoury filo pie) 24
  bourekia me revithia 119
  bourgiourdi (roast feta en papillote) 62
  fig saganaki 204
  goat on grilled pita with purslane and grated cheese 187
  heli spetsiotiko (pot-roasted eel, Spetses style) 155
  kaseropita with basil 64–5
  lamb baked in paper 147
  lobster plaki with a fig and manouri salad 97
  manouri and pistachio ice cream 211
  mussels and clams on the barbecue 163
  omelette with honey and sesame seeds 26
  pan-seared fig, manouri and pastrouma sandwich 184
  pasta triangles 171
  prawns with apricot feta salad 87
  run out of flour for bread? 121
  sfakianopites (cheese pies) 207
  tirokeftedes 181
  trahana soup with manouri cheese 14–15
  yoghurt-based cheese with rosemary 133
cherries 13
  cherry butter 12
chicken: borlotti beans with grilled chicken wings 54
  chicken soup trahana 137
  okra and chicken filo rolls 42–3
  pressed cockerel 162
chicken livers, fragrant nut rice with 76–7
chickpeas: leg of lamb with chickpeas 80
  revithosoupa (chickpea soup) 116
chilli peppers: bourgiourdi (roast feta en papillote) 62
  spicy tomato and walnut dressing 129
clams: mussels and clams on the barbecue 163
  razor clams flambéed with ouzo 99
cockerel, pressed 162
cockles, lemony rice pilaff with 88
coffee 19
crab meat on grilled sourdough bread 93–4
Crete 13, 14, 50
cuttlefish, stuffed 108–9
Cyclades 59, 74

D
dips: koliosalata (smoked mackerel salad dip) 70
  parsley and onion dip 164
dolmades, Cretan 32–3
dried fruit: apples for Freddie 127

E
eel: heli spetsiotiko (pot-roasted eel, Spetses style) 155
eggs: fried eggs with chopped throumbes olives and tomatoes 16
  froutalia 125
  omelette with honey and sesame seeds 26
  salad of green cracked olives, pistachios, egg and sesame seeds 192
  trahana 138
elephant bean casserole 122
Evia (Euboia) 96

F
fennel: elephant bean casserole with orange and fennel 122
  fennel rissoles from Serifos 74
  grilled whole fish with fennel 167
  sea urchins, avgotaraho, fennel and kohlrabi salad 91
feta see cheese
figs: fig saganaki 204
  lobster plaki with a fig and manouri salad 97
  pan-seared fig, manouri and pastrouma sandwich 184
fish: grilled whole fish with fennel 167
  see also mackerel, red mullet etc
fisherman's relish – rabbit spread 158
flomaria (hilopites) with milk-roasted baby goat 142–3
Folegandros 75
fritters: savoury loukoumades 183
froutalia 125
fruit: after-lunch Aegean mint fruit salad 196
  watermelon and other fruits 201
  see also apples, figs etc

G
goat: flomaria (hilopites) with milk-roasted baby goat 142–3
  goat on grilled pita with purslane and grated cheese 187
  grilled kid with green tomatoes 141
  sausages 124
gourounaki kokkinisto (pork casserole) 130–1

H
Halki 144
heli spetsiotiko (pot-roasted eel, Spetses style) 155
hilopites with milk-roasted baby goat 142–3
honey: pasteli 212

I
ice cream: green tomato 210
  manouri and pistachio 211
Ionian Islands 203

J
jelly: strawberry pelte 203

K
Kalymnos 160
Karpathos 108
kaseropita with basil 64–5
Kato Koufonisi 157
Kea 212
kid see goat
kohlrabi: sea urchins, avgotaraho, fennel and kohlrabi salad 91
koliosalata (smoked mackerel salad dip) 70
Kythira 55

L
lamb: lamb baked in paper 147
  lamb mastella 136
  leg of lamb with chickpeas and seasonal vegetables 80
lamb tongues, grilled artichokes and 188
leeks: lemony rice pilaff with cockles 88
  sausages 124
lentils: lentils with a touch of spices from Mytilini 120
  warm salad of red mullet, lentils, egg and parsley 178
Lesvos 68, 79, 182
lime juice: frozen ouzo mojito 215
liver see calves' liver, chicken livers
lobster plaki with a fig and manouri salad 97
loukoumades, savoury 183

M
mackerel: koliosalata (smoked mackerel salad dip) 70
  mackerel fillets with caper leaves and fresh dill 73
mandarin, olive and sweet red onion salad 161
manouri see cheese
marida see whitebait
mastic, honey and peach avgolemono drink 20
meat, potting 186
  see also goat, lamb etc
milk: run out of flour for bread? 121
  trahana 138
  yoghurt 10
  yoghurt-based cheese with rosemary 133
mojito, frozen ouzo 215
mosxaraki kapama (veal stew) 145
mussels: mussel soup 98

mussels and clams on the barbecue 163

N

noodles: flomaria (hilopites) with milk-roasted baby goat 142–3

O

octopus with rosemary and garlic 105
oil, basil 70
okra and chicken filo rolls 42–3
olive oil: anchovy drippings 182
olives: fried eggs with chopped throumbes olives and tomatoes 16
  mandarin, olive and sweet red onion salad 161
  salad of green cracked olives, pistachios, egg and sesame seeds 192
  omelette with honey and sesame seeds 26
onions: button onion bulbs with dill 61
  heli spetsiotiko (pot-roasted eel, Spetses style) 155
  mandarin, olive and sweet red onion salad 161
  parsley and onion dip 164
  rabbit and peas hotpot 144
  tirokeftedes 181
orange: elephant bean casserole with orange and fennel 122
  mandarin, olive and sweet red onion salad 161
  mosxaraki kapama (veal stew) 145
ouzo: frozen ouzo mojito 215

P

parsley and onion dip 164
pasta 170
  pasta triangles 171
  trahana soup 14–15
pasteli 212
pastrouma (cured beef): pan-seared fig, manouri and pastrouma sandwich 184
  razor clams flambéed with ouzo 99
peaches: grilled trahana and peach with yoghurt 206
  peach-watermelon preserve 11
  warm mastic, honey and peach avgolemono drink 20
peas: lamb baked in paper 147
  rabbit and peas hotpot 144
peppers: froutalia 125
  mussels and clams on the barbecue 163
  Santorinian oven-roasted tomatoes and peppers 36
pies: bougatsa 24
  bourekia me revithia 119
  kaseropita with basil 64–5
  okra and chicken filo rolls 42–3

sfakianopites (cheese pies) 207
pilaf: bulgur, walnut and spinach pilaf 39
  lemony rice pilaff with cockles 88
pine nuts: fragrant nut rice with chicken livers 76–7
pinna, pan-seared 110
pistachios 13
  fragrant nut rice with chicken livers 76–7
  manouri and pistachio ice cream 211
  roasted pistachio biscuits 12
  salad of green cracked olives, pistachios, egg and sesame seeds 192
pomegranates: bulgur wheat salad 179
pork: gourounaki kokkinisto (pork casserole) 130–1
  roast pork loin on the bone 148
  sausages 124
potatoes: fried squid with a salad of artichoke, potato and anchovy 101–2
  froutalia 125
  lamb baked in paper 147
  warm salad of potatoes 75
prawns: prawns with apricot feta salad 87
  Symi shrimps with tomatoes and caper leaves 51
preserve, peach-watermelon 11
purslane: borlotti beans with grilled chicken wings and purslane 54
  goat on grilled pita with purslane and grated cheese 187

R

rabbit: fisherman's relish – rabbit spread 158
  rabbit and peas hotpot 144
razor clams flambéed with ouzo 99
red mullet, warm salad of 178
revithosoupa (chickpea soup) 116
rice: fragrant nut rice with chicken livers 76–7
  lemony rice pilaff with cockles 88
  stuffed cuttlefish 108–9
rissoles, fennel 74

S

salads: bulgur wheat salad 179
  fried squid with a salad of artichoke, potato and anchovy 101–2
  lobster plaki with a fig and manouri salad 97
  mandarin, olive and sweet red onion salad 161
  mixed green salad from Lesvos 69
  prawns with apricot feta salad 87
  salad of green cracked

olives, pistachios, egg and sesame seeds 192
  sea urchins, avgotaraho, fennel and kohlrabi salad 91
  seasonal salad 148
  warm salad of potatoes 75
  warm salad of red mullet, lentils, egg and parsley 178
Samothraki 40, 168
Santorini 36–7
sauces: anchovy drippings 182
sausages 124
  froutalia 125
sea bream, barbecued 176
sea urchins, avgotaraho, fennel and kohlrabi salad 90–1
semolina: pasta triangles 171
Serifos 74
sesame seeds: pasteli 212
Sfakia 207
sfakianopites (cheese pies) 207
shallot vinaigrette 79
shrimps see prawns
Sifnos 136
smoked mackerel salad dip 70
snails with stoneground wheat 50
soups: chicken soup trahana 137
  mussel soup 98
  revithosoupa (chickpea soup) 116
  trahana soup 14–15
Spetses 155
spinach: bulgur, walnut and spinach pilaf 39
squid with a salad of artichoke, potato and anchovy 101–2
strawberries: strawberry pelte 203
  trahana soup with 14–15
sweetbreads: broad beans with braised lamb sweetbreads 41
Symi shrimps 51
Syros 145, 148, 164

T

tea 'the Greek way' 25
terrines: pressed cockerel 162
tirokeftedes (deep-fried cheese balls) 181
tomatoes 37
  anchovy fillets with Santorinian capers and dill 79
  aubergine stifado 49
  Cretan snails with stoneground wheat 50
  flomaria (hilopites) with milk-roasted baby goat 142–3
  fried eggs with chopped throumbes olives and tomatoes 16
  gourounaki kokkinisto (pork casserole) 130–1
  green tomato ice cream 210
  grilled kid with green tomatoes 141
  heli spetsiotiko (pot-roasted eel, Spetses style) 155

lobster plaki with a fig and manouri salad 97
  mussel soup 98
  roast pork loin on the bone 148
  roast tomatoes 39
  Santorini oven-roasted tomatoes and peppers 36
  spicy tomato and walnut dressing 129
  sweet tomato perasti with aged red wine vinegar 128
  Symi shrimps with tomatoes and caper leaves 51
tongues: grilled artichokes and lamb tongues 188
trahana 138
  chicken soup trahana 137
  Cretan dolmades 32–3
  grilled trahana and peach with yoghurt 206
  run out of flour for bread? 121
  trahana soup 14–15

V

veal: mosxaraki kapama (veal stew) 145
  slow-cooked veal cheeks 168
vegetables: elephant bean casserole with orange and fennel 122
  leg of lamb with chickpeas, seasonal vegetables and mint 80
  mosxaraki kapama (veal stew) 145
  slow-cooked veal cheeks 168
  see also peppers, tomatoes etc
vinaigrette, shallot 79
vine leaves: bonito and Santorinian spring vine leaf rolls 156
  Cretan dolmades 32–3

W

walnuts: caramelised walnuts 196
  spicy tomato and walnut dressing 129
watercress, celeriac roots and 159
watermelon 200
  peach-watermelon preserve 11
  watermelon and other fruits 201
wheat: Cretan snails with stoneground wheat 50
  see also bulgur wheat
whitebait (marida): beetroot with fried marida 45

Y

yoghurt 10
  Greek yoghurt with roasted pistachio biscuits and cherry butter spread 12
  run out of flour for bread? 121
  trahana 138
  yoghurt-based cheese with rosemary 133

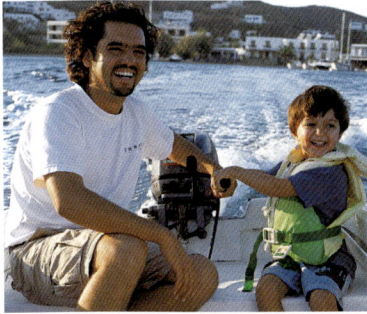

Acknowledgements

First, thanks to Panos Manuelidis: a personality off the radar, off the sailing map and free-spirited. Over my last 8 years I have adopted a few of his enthusiasms. His company, **Odysea**, donated the following ingredients to us:

Aged Corinthian vinegar
Greek feta cheese (the feta was produced by Roussas, *see* the Glossary for details)
Olives (hidden in the hillsides of the Greek island of Evia, you find Rovies, a single estate grove, where they still harvest their olives by hand)
Graviera Kritis and Savoury Graviera
Manouri
Kasseri
Kefalotiri
Greek strained yoghurt
Tomato perasti
Wild thyme and pine and fir tree honeys (produced by a local beekeeper in a small village of Ilias in central Greece. The bees during the months of spring and summer enjoy their holidays in areas where wild thyme, herbs and conifers grow in abundance)
Vine leaves
Roasted red peppers (sweet, whole flame-roasted red peppers from the Florina region of Northern Greece)
Iliada PDO Kalamata Extra Virgin Olive Oil (produced from Koroneiki olives, which are grown in the Kalamata region of the Peloponnesus, this has a distinctive grassy green colour and herby flavour with a very pleasant peppery aftertaste. Since 1997, the extra virgin olive oil produced in this area has been awarded PDO status.
Traditional flat pita bread
Greek dried pulses, nuts and Krokos saffron (Elephant beans, peeled chickpeas, small lentils, short-grain rice, fava beans, pine nuts, walnut kernels, roasted salted pistachios in their shells, roasted almonds, Corinthian sultanas and red saffron from Kozani)

Also thanks to **Papagiannakos Estate:** Vasili, your Savatiano was a big hit with the group; it was enjoyed by all. Very clean and crisp. The wine went perfectly with 90% of what we ate and was especially enjoyable as an aperitif. Your Cabernet Sauvignon was also such a beautiful wine, with bags of fruit, excellent balance, not too much oak; it was particularly good with the kid and also the lamb.

Dimitri Seitanidi and **TACOM Wine Services**, thank you for:
1 Raptis Estate, Messogios Rose: Very nice easy drinking, slightly on the sweet side. Good rich pink colour and lots of strawberries on the nose. We all enjoyed it as our aperitif wines and we probably could have got through twice as much
2 Papantonis Megan Agan: Quite dry, this benefited from being open for a little time, and we enjoyed it with our pork, with which it made a very happy bedfellow!! It had a very nice earthiness about it, giving a really good rustic flavour.
3 Nasiakos Mantinia: A very nice example of a moscofiliero, beautifully green and very crisp and flinty, which also delivered on flavour. It was perfect with the dentex that we ate and also the squid and sea urchins.

A special thanks to **Denton's Catering Equipment** and Michael Nunn for your generosity, which made the cooking on board the boat so much easier.

My personal thanks to Paloma, Lori, Stephanie, Patrick, Arthur and Alexandra for the splendid company and help during the sailing trip. I am sorry for the ones who felt so seasick!

I have come so close to the end of this book and I have spoken about almost everything, but not about the photography. Shame? It was easy to get lost in Jason's box, especially when he was staring at the turbulent horizon. Jason, I beg your pardon for forgetting to tell you that the Eastern Aegean is a windy country. Your attitude towards 'let's sail with no plans' made me recognize the street you were seeking in your lens's magenta, which, in a circular loop, brought the purple enclosure of our silent thoughts that were your real destination. Jason, once more, I enjoyed the oracle of the intriguing bond between our characters, as dissimilar as the fig and the artichoke!

The Meltemi, blowing mostly NNE with a constant Force 7 to 8 was with us throughout the whole of our last week and Ali Oskurak, our skipper, handled not only the swirling and twirling strong winds as one of his daily routines but his beautiful boat too, Grandi, with his magnificently trained three elves, Hakan Uguz, Zafer Sanlı and Öztürk Çat, his crew. I have sailed with Ali a few times and what makes him so special is that he never argues with the Greeks, he simply kills them with his politeness! Tussock, the company behind the Grandi has kept all its 9 boats seaworthy, very comfortable to live in and with very good crews, hence they cater very well for individually tailored holidays. I run some of my cookery workshops on Tussock's boats.

Summoning up my memories, let's write the epilogue from its foundations:
1 Hakan, the chef on board, who kept awake throughout the whole trip against my chopping board's swaying, and the Grandi's rocking, while at the same time he was trading all captain's sailing instructions at every solstice and equinox.
2 Freddie, as soon as you set foot on the boat, everything that I had planned was forgotten; the shortest point between two thoughts was never a straight line but a zigzag of tortuous routes, which all led to the island with Esmeralda.
3 Chantal, your harmonious presence with its divine origin that over the years has aroused no controversy, but the map of the Magellan Cloud where we revolve, travel and dream.
4 Mum, thank you for your moments around the kitchen table with your splendid cooking smells.
5 Jane and Helen, for harbouring all my 'Greekeries', apart from hacking away all the pictures from the goat sacrifice in Agathonisi.
6 Lucy, for the melody of subsuming the rhythm of all the photographs and text on your chopping board, a slow and elaborate action, which kept it awake for weeks.
7 Lewis, for the velvet smoothness of your editing of my text and thank you for bringing all of us to the invisible rolling coast of sweets!